I0006722

China's Cyber Warfare

China's Cyber Warfare

The Evolution of Strategic Doctrine

Jason R. Fritz

LEXINGTON BOOKS
Lanham • Boulder • New York • London

Published by Lexington Books
An imprint of The Rowman & Littlefield Publishing Group, Inc.
4501 Forbes Boulevard, Suite 200, Lanham, Maryland 20706
www.rowman.com

Unit A, Whitacre Mews, 26-34 Stannary Street, London SE11 4AB

Copyright © 2017 by Lexington Books

All rights reserved. No part of this book may be reproduced in any form or by any
electronic or mechanical means, including information storage and retrieval systems,
without written permission from the publisher, except by a reviewer who may quote
passages in a review.

British Library Cataloguing in Publication Information Available

Library of Congress Cataloging-in-Publication Data
The hardback edition of this book was previously catalogued by the Library of Congress as follows:

Names: Fritz, Jason R., 1977- author.
Title: China's cyber warfare : the evolution of strategic doctrine / Jason R. Fritz.
Description: Lanham, MD : Lexington Books, [2017] | Includes bibliographical references and index.
Identifiers: LCCN 2016055925 (print) | LCCN 2016056128 (ebook)
Subjects: LCSH: Cyberspace operations (Military science)—China. | Information warfare—China. |
    Military doctrine—China.
Classification: LCC UA835 .F85 2017 (print) | LCC UA835 (ebook) | DDC 355.3/43—dc23
LC record available at https://lccn.loc.gov/2016055925

ISBN 9781498537070 (cloth : alk. paper)
ISBN 9781498537094 (pbk. : alk. paper)
ISBN 9781498537087 (electronic)

∞™ The paper used in this publication meets the minimum requirements of American
National Standard for Information Sciences Permanence of Paper for Printed Library
Materials, ANSI/NISO Z39.48-1992.

Printed in the United States of America

Contents

Introduction

The People's Republic of China (PRC, China) acquired a singular notoriety in the years 1998 to 2014. This was the period when multi-state accusations of computer network intrusions by China gained prominence, while existing open source knowledge on this topic remained superficial and misinformed. Cybercrime, including state-sponsored cybercrime, was estimated in 2014 to cost the global economy $400 billion annually, and China itself has been one of the largest victims (McAfee 2014). Concurrently, China operates one of the world's most sophisticated Internet censorship and monitoring services, and has the world's largest online population at 632 million, with 527 million accessing the Internet through mobile devices (USCC Annual Report 2014). The Internet has become vital to the functioning of a modern society. China's economy has rapidly risen to the world's second largest, and along with it, the People's Liberation Army (PLA) has accelerated modernization, seeking the capability of "winning local wars under the conditions of informationization" (xìnxī huà; 信息化) (Information Office of the State Council of the People's Republic of China 2013). Moreover, cyberspace has become a recognized domain of warfare, in addition to land, sea, air and outer space. The 2013 edition of *The Science of Military Strategy*, produced by the PLA's Academy of Military Science, concedes that the PLA possesses defensive and offensive cyber units. All of these factors, combined with Beijing's growing assertiveness in international relations, have generated increased interest in China's intentions. Such developments are readily chronicled but inadequately conceptualized within the larger framework of China's military thought. Reasons for such a shortcoming in the literature include the fast pace of developments in cyber security and a lack of transparency in Chinese military affairs. This may be due to genuine disagreement within the government, or intentional to hide motives. Much of the writings published by the PRC are contradictory or ambiguous, using modern and ancient foundations, while being disseminated by varied sources. Unlike the United States or the North Atlantic Treaty Organization (NATO), and despite publication of defense white papers every two years, China is not transparent in its strategic doctrine. This needs to be inferred. Hence, this book will pursue the question: *What is the strategic doctrine to emerge from China's development of cyber warfare? The method employed, as elaborated below, is through a conceptual and historical investigation.*

ORIGINALITY AND JUSTIFICATION

Understanding China's cyber warfare (wǎnguò zhàn; 网络战) doctrine fills a significant gap in the knowledge base of Chinese military thought. Beijing has never published a formal cyber warfare doctrine (Gady 2015; Mulvenon 1999; USCC Annual Report 2012). Chinese military journals and newspapers provide valuable insights, and a number of external "open-source researchers" have pieced together some of China's capabilities, goals, and strategies (USCC Annual Report 2009). However, these lack detail, specifically in issue area, and they are not a "definitive work representing a concise and coherent view" (Krekel, Adams, and Bakos 2012; Thomas 2004, 3). Despite these inadequacies, the twin phenomena of China's military modernization and the ubiquity of computer networks in the performance of modern weapon systems have attracted global concern. China's increased military prowess will alter the balance of power in the Asia Pacific region, enabling China to exert greater influence on contentious issues such as maritime borders, resource allocation, Taiwan re-unification, and the United States' military presence in the Western Pacific. Advancements in space-based capabilities and the PLA Navy (PLAN) have provided China with limited force projection beyond the Asia Pacific region, yet in the cyber realm it already possesses force projection on a global scale. Additionally, the Internet can enhance its competitiveness in the other domains through technology transfer, propaganda, and eroding a foreign military's capabilities. China is cited recurringly as taking a holistic approach to cyber warfare. In keeping with a holistic approach, this book comprehensively examines open source information which is applicable to Chinese cyber warfare. This includes examining historical developments that help clarify the prevailing situation, in contrast to a tendency in the cyber warfare literature to focus on the latest cyber incidents or revelations as though they were new occurrences when in actuality there had been precedents and recurring patterns. For instance, it is common to read that the 2007 cyberattacks on Estonian were "the world's first cyber conflict" and the Stuxnet worm was "the first cyber weapon"; however, this book will show these statements are not true. Similarly, the phrases "electronic Pearl Harbor" and "watershed moment" have been attached to a wide range of cyber incidents, yet they were not the first such incidents and they did not result in drastic policy changes.

Conceptualizing China's cyber warfare doctrine requires clarifying and consolidating an overabundance of vague and overlapping terms. To accomplish this, this book has created a three-branch[1] model of cyber warfare based on core distinctions (see the *Definitions* subsection below for further examination of terminology). This model provides a simplified structure for all-inclusive analysis. Part of China's cyber doctrine can be inferred by examining the goals it is seeking and ascertaining how those could be met through advancements in information and communication technology (ICT). This book evaluates China's state-level goals, rather than the mission statements of a few frequently changing cyber-related entities. Secondly, it examines all that cyber warfare includes, or can include, rather than what one definition, or a handful of different definitions, include. Streamlined categorizing will aid the efficiency of this research by bringing order to a diverse range of topics and allowing duplicate concepts or incidents to be merged. This could also aid future researchers in the same manner, or it could allow them to limit the scope of their research by focusing on a particular aspect within cyber warfare. Improved conceptualization will allow the examination of cyber warfare to advance, rather than become trapped in a continual process of renaming and rediscovering. In particular, it might reveal novel ideas in past research which were initially displaced by the volume of inquiry in the emerging field of cyber warfare or open the way for the identification of new ideas, weaknesses, and trends. This categorization can also limit media and public misunderstanding, and it can provide more accurate threat assessments (discussed further in chapter 1, under Computer Network Attack). This book will use this model in order to identify the Chinese government entities responsible for developing cyber warfare, in all its aspects, and their relationship to one another. Doing so facilitates a new perspective, one which might reveal ways to improve the organizational structure, or reveal entities which would mutually benefit through collaboration. Lastly, all of these factors work toward increased transparency, which can advance the cause of cooperation, understanding, and trust.

BACKGROUND

An examination of China's goals and the development path it has taken will show the aims of the military and the strategic thought behind its actions. The PLA has tried to transform itself from a land-based power to a comparatively smaller, more mobile, high-tech power that is capable of reaching beyond its borders. During the 1980s paramount leader Deng Xiaoping pushed for quality over quantity, and the military was reduced by one million personnel. In 1993, President Jiang Zemin officially announced a Revolution in Military Affairs (RMA) as part of the national military strategy for modernization. RMA, which is discussed in detail in chapter 4, is a theory on the future of warfare connected to technological changes and associated organizational changes. Careful observation of U.S. involvement in the Kosovo, Afghanistan, and Iraqi wars, furthered China's interest in Net-Centric Warfare (discussed in chapter 4) and asymmetric warfare, the former successfully used by the United States, and the latter successfully used against the United States. At the turn of the century, the bulk of China's traditional military force retained 1950s to 1970s era technology, which was imported and reverse engineered from the then Soviet Union. China has sought to modernize this force. The size of China's traditional force will shrink, as fewer numbers are needed when new technology is introduced. China seeks to maintain domestic and regional stability while developing its economic, military, technological, scientific, and soft power. It also seeks a balance between military and economic development, believing they are mutually dependent. Beijing maintains its One China Policy in relation to Taiwan, and claims sovereignty over the Parcel and Spratly islands and adjacent waterways.

Deng Xiaoping, representing second-generation leadership after Mao, sought to avoid international responsibilities and limitations, as they could slow down development of the military and economy. The third-generation leadership of Jiang Zemin did look outward, promoting a multipolar world in the face of the post–Cold War unipolarity under the United States, just as fourth-generation leader Hu Jintao promoted the ideology of a Harmonious World (Héxié shìjiè, 和谐世界) which places more emphasis on international relations. The successor to Hu Jintao, President Xi Jinping, emphasized the "Chinese Dream" that included a diversity of culture in the global order. However, the PRC continues to avoid a definitive posture in terms of international commitment through concepts of non-interference, diversity, and equality. It compares itself to other states through Comprehensive National Power (CNP—Zònghé guólì, 综合国力), using qualitative and quantitative values, and not accepting traditional Western categorizations. For example, China includes the economy, soft power, and domestic stability as factors of CNP. This is significant in that it shows a correlativity which holds relevance for cyber warfare. Under CNP the economy, soft power, and domestic stability can be seen as militarily relevant matters. Further, maintaining the status quo in regards to Taiwan and the Spratly Islands may not be China's long-term intention, but rather a way to stall efforts while it builds up military strength, strength which can include economic and international influence.

Despite not wanting to become embroiled in concrete commitments to military strategy, Chinese leaders cannot ignore the interconnectedness of the modern world, and they have realized the necessity of international cooperation. Notably, the need for resources has fuelled China's global presence. The PRC is the world's largest net importer of petroleum. As the country's economy grows and the middle class expands, the demand for fossil fuel resources will continue to grow. This creates a requirement for sound international relations with exporting nations and the need for security of transportation routes, such as the Strait of Malacca linking the Indian Ocean and the South China Sea and expanding networks of overland pipelines. The politics and military affairs of the states involved are thus relevant to China's concerns. Competition with the United States for these resources has often led to China making agreements with nations the United States opposes on several points, such as Angola, Chad, Egypt, Indonesia, Iran, Kazakhstan, Nigeria, Oman, Sudan, Venezuela, and Yemen.

In these suppliers Beijing finds less competition for resource access. However, the result is often international criticism of China as these states may be violating human rights or supporting terrorism. Moreover, Beijing's methods of befriending these exporters comes into question, especially in regards to arms being traded or availability of finance which may be supporting controversial policies. China currently lacks the power projection to protect critical sea lanes from disruption or to significantly deter international criticism. Crucial to extended power projection is a "blue water" navy which would benefit from online technology transfer and the further development of Command, Control, Communications, Computers, Intelligence, Surveillance and Reconnaissance (C4ISR). Online Psychological Operations (PSYOPS) and media warfare would enhance China's soft power. Beijing believes that economic growth is critical to military development; economic growth increases energy demand, which in turn creates a greater military demand; thus the two are mutually reinforcing. Certainly, Xi Jinping's "Silk Road Economic Belt" and the "21st-Century Maritime Silk Road" initiative, termed "One Belt, One Road," serves both an economic and security function. This large-scale infrastructure development project assists not only regional economies but also China's quest for resources and the security of their transportation routes. Moreover, the ability to reach markets by both land and sea routes is also secured.

Ensuring the survival of the Communist Party of China (CPC) shapes China's strategic outlook. In order to bolster domestic support for policies, nationalism has been emphasized over communist ideology. This can be seen with government-organized protests against Japan, over visits by Japanese leaders to World War II shrines and protests against the publishing of Japanese school textbooks which downplay Japan's atrocities against the Chinese. These protests tend to coincide with other strategic interests, such as territorial disputes in the East China Sea which, with the notable exception of the highly publicized arrest by Japan of a Chinese captain in September 2010 off the disputed Senkaku/Diaoyu Islands, are often unbeknownst to the casual observer or participant. The mobilization of nationalism was also evident during the holding of a U.S. reconnaissance aircraft in 2001, and the mistaken NATO bombing of the Chinese Embassy in Belgrade in 1999. The 2008 Beijing Olympics further demonstrated how China could garner national support in the face of corruption, environmental degradation, and a widening wealth gap. These events highlight a strategic value in public manipulation through nationalism; one that is interconnected with military affairs, and one which is increasingly turning to online assets.

Ensuring the survival of the CPC comes not only from the above strategic nationalist discourse. It intimately concerns the PLA. "Consolidate the ruling status of the CPC" is the first of the missions entrusted to the PLA, which is a "party-army" in that it brought the CPC to power and swears allegiance to the Party. Several conclusions can be drawn from the status of the PLA. China is committed to modernizing its military, in part through the purchase or illicit acquisition of foreign technology and subsequently reverse engineering that technology so it can be produced domestically. Defence reforms in support of modernization have included favoring quality over quantity by decreasing the PLA's size: from 4 million to 2.25 million personnel. The PLA's weaponry often lags one or two generations behind that of Western military powers. However, the total force base still poses a significant deterrent, and establishes China as a dominant power within the Asia-Pacific Region. Other than deployments to the Gulf of Aden on anti-piracy missions, China lacks force projection beyond its region, primarily due to the lack of a blue water navy and aircraft carrier fleet,[2] but also due to limits in missile technology and air-defense penetration, and opposition by foreign powers such as the United States. China seeks to become self-sufficient in many of these key capabilities. Once it has leapfrogged and is no longer trying to catch up, in accordance with policy reflected in China's defense white papers, the PRC will no longer need such wide-scale technology transfer, and it will possess the might to shape the international system, rather than be bound by one that was created by foreign powers.

OPEN SOURCE MATERIAL

Although commonly regarded to as secret or scant, knowledge relating to Chinese cyber warfare may be found in a large body of open source material. This includes articles from defense contractors, government organizations, information technology (IT) security firms, military journals, news agencies, think tanks, universities, and other non-government organizations (NGOs). Despite being disjointed, superficial, and misinformed, Chinese and Western news articles do provide information towards inferring China's cyber doctrine that is not available anywhere else and therefore cannot be ignored. This is a consequence of cyber warfare being an emerging area of study that is quickly changing, a lack of transparency, and a reluctance of businesses and governments to make direct accusations of aggression. Therefore, this mass collection of smaller (in size or scope) articles is necessary. For example, foreign accusations of Chinese computer network intrusions provide insight into what is possible through cyber warfare and the direction in which it may be headed.

Further, even if these accusations prove to be unfounded, they illustrate a need for cyber defense, diplomacy, law, and transparency. Research reports conducted by think tanks provide a greater depth of information than news agencies, although they often focus on a single aspect of China or cyber warfare, such as terrorism or computer network exploitation. Further, these reports often lack a common cyber-terminology foundation (discussed below under the subheading *Definitions*), which can result in disjointed explorations, repetition, and less depth than what is possible with a consolidated terminology-aiding organization. IT security firms, meanwhile, provide valuable information through the examination of advanced persistent threats (discussed in chapter 2). However, these focus on a particular incident or group of incidents, rather than an overarching examination of statewide capability, goals, and strategy. Understandably, these commercial entities are also reluctant to accuse China directly of wrongdoing, and they lack expertise in the field of international relations.

Moving into more substantive literature, this research examines key government-affiliated documents. Firstly, two significant collections are the U.S. Department of Defense's annual reports to Congress on the military power of the People's Republic of China, and the US-China Economic and Security Review Commission's (USCC) annual reports to Congress, as well as the USCC's regular hearings, roundtables, and special reports. These reports provide a review of each year's open source knowledge on China's cyber developments, and they offer a glimpse of closed source intelligence. They also provide original information through extensive field research, interviews, testimony, and translations; and they reveal how this fits in with China's overall defense doctrine. These U.S. reports are compared and contrasted with key Chinese government reports, including China's bi-annual National Defense white papers, China's Five Year Plans, The National Computer Network Emergency Response Technical Team/Coordination Center Annual Reports, and a number of singular white papers like *The Internet in China* (2010). While these white papers are lacking in transparency and depth, they do confirm issues mentioned in the open source material, such the Chinese government's commitment to expanding cyber capabilities and training. They also provide insight into possible cultural aspects of Chinese cyber warfare that do not appear in Western reports. Cultural considerations include the infusion of Confucian ideas, China's long history, the Century of Humiliation, Harmonious Society, and the teachings of Mao Zedong and Sun Zi (Sun Tzu), which are discussed below.

A number of essential texts pertaining to China's development of cyber warfare doctrine will be referred to throughout this book. Prominent Western authors include James Andrew Lewis, James Mulvenon, Larry M. Wortzel, Martin C. Libicki, Richard A. Clarke, Scott Henderson, and Timothy L. Thomas. Timothy L. Thomas in particular produced a trilogy of books—*Dragon Bytes*, *Decoding the Virtual Dragon*, and *The Dragon's Quantum Leap*—replete with translations and analysis of a wide range of influential PLA texts pertaining to cyber warfare. While these books are indispensable, they do suffer from hindrances present in the Chinese source material. Chinese authors, like their Western counterparts, have not reached consensus on terminology, and this can stymie a more sophisticated treatment of the cyber warfare topic. That prominent Chinese authors are still developing these concepts is in itself revealing. However, there is a need to coalesce all of these thoughts and move forward. Similarly, two reputable Chinese books on military strategy—*The Science of Campaigns* and *The Science of Military Strategy*—are also said to "lack specificity on cyber operations" (USCC Annual Report 2012). These books do, however, demonstrate the importance given to cyber warfare, and they advocate developing a holistic approach which includes all three branches discussed below. Going even further, *Unrestricted Warfare*, a book by two PLA senior colonels, Qiao Liang and Wang Xiangsui, claims that warfare is no longer strictly a military operation, and that the battlefield no longer has boundaries. *Unrestricted Warfare* was published by the PLA Literature and Arts Publishing House in Beijing in February 1999. According to the Foreign Broadcast Information Service (FBIS) translation editor, the book "was endorsed by at least some elements of the PLA leadership" and an interview with one of the authors was published in the Communist Party of China (CPC) Youth League's official daily newspaper on June 28, 1999. Thus while the book is not entirely backed by the PLA, especially the older generation, it is not without merit as a source. To use the "half empty, half full" glass analogy, it may seem inadequate in terms of its ambiguous status. Yet it does have some official backing and hence a degree of legitimacy as a document assisting analysis, not only on the PLA's likely developmental trajectory but also how asymmetric tactics against a superior hi-tech military might be employed. The authors of *Unrestricted Warfare* assert that war has not disappeared, but its appearance has changed and its complexity has increased (Liang and Xiangsui 1999).

China's strategy can adapt to modern technology without losing some of its long-standing traditions, such as its distinctive style and modes of thinking pertaining to national security. The concept of the weak overcoming the strong in Daoist philosophy, for example, has renewed relevance in computer network exploitation and anti-access area denial (discussed in chapter 2 and chapter 4, respectively). Two seminal authors in China's history are Sun Zi and Mao Zedong. Sun Zi promoted deception in warfare and the strategy of winning a war before it reached the physical battlefield. There does not appear to be a definitive text which comprehensively applies Sun Zi's proverbs to cyber warfare, but deception and unconventional warfare can flourish in the cyber realm. Sun Zi's teachings work in harmony with online characteristics such as anonymity, dual use, high speed, a relatively low cost and low entry-level, the removal of geographical distance, and an underdeveloped legal framework to guide responses to attacks. Mao Zedong's teachings also have renewed importance within cyber warfare where virtual space provides increased means and depth for conducting guerrilla warfare which is a component of People's War. People's War is no longer restricted to the mainland, as the Internet provides global reach, and China has the world's largest online population. Moreover, the Internet has increased the ability to mobilize the masses and facilitated their capacity to participate in defense. China's national defense white papers in 2008 and 2010 both call for the armed forces to develop new "strategies and tactics for conducting people's war" based on information systems (Information Office of the State Council of the People's Republic of China 2008; Information Office of the State Council of the People's Republic of China 2013). Cataloguing software exploits, cultivating hacker groups, mapping foreign networks, and maintaining a covert presence within foreign networks (discussed in chapter 2) all fall within Mao's "active defence" strategy. Engaging in offensive action for the purpose of self-defense—which is a feature of active defense—is also illustrated in information operations (discussed in chapter 3). The application of both people's war and active defense in the cyber realm have been examined by Chinese theorists since the late 1980s (Thomas 2004, 40, 44–45, 48–49).

The use of asymmetric warfare forms a common thread between Sun Zi and Mao Zedong. A militarily disadvantaged China can use its special advantages, or effectively exploit its enemy's particular weakness, to prevail in the event of conflict. Should conflict occur over Taiwan declaring independence, for example, China is significantly outmatched by the military force the United States can bring to bear. However, the United States relies heavily on computer networks for navigation, reconnaissance, communication, and the logistical aspects of force deployment. By developing cyber warfare capabilities, the PRC can exploit a weakness within the U.S. arsenal in order to increase its combat effectiveness, such as delaying or distracting U.S. force deployment. Cyber warfare also acts as a deterrent to conflict, in part because the United States relies on computer networks for its economy, infrastructure, and daily life; though for the deterrent to be credible it would require an acknowledged offensive capability, as occurs in the nuclear realm. Possessing such a deterrent would buy time for China to advance to higher standards in traditional military power. Cyber warfare can debilitate an opponent, act as a force multiplier, and equalize opportunities against adversaries that are superior in traditional warfare. While Beijing recognizes the need for international cooperation, it remains cautious. Its "Hundred Years of Humiliation" are part of the national discourse. Colonialism by Western powers, Japanese occupation in World War II, and Chinese intervention in the Korean War and involvement in the Vietnam War were not the only experiences to influence Chinese security thinking. Border conflicts with India in 1962, the Soviet Union in 1969, and Vietnam in 1979 were to follow. Despite China's long history, these events are of special note as they are within living memory, and they marked the founding and maturation of the CPC's rule.

China's cyber warfare doctrine incorporates all three levels of warfare—strategic, operational, and tactical. In general, the strategic level deals with state-level goals and decisions, such as national security, foreign policy, and the decision to go to war. This level can be further divided into "grand strategic" and "military strategic," with the latter being focused primarily on the conduct of war. The operational level relates to strategy within a theater of conflict, and can sometimes be distinguished by the application of code names to the operations. Finally, tactical-level strategy pertains to individual units and direct engagement with the enemy; leadership and morale play a large role at this scale. Cyber warfare, however, complicates this segmented approach to addressing problems. The ability to act from great distances at high speed, and the increasingly non-linear dissemination of information and command, has caused time-space compression. Additionally, it has blurred the separation between civilian and military participants and targets. By removing geographic barriers and traditional roles, the tactical level can now impact the operational and strategic levels, such as a hacker disrupting force deployment logistics networks or targeting large-scale financial systems. A complete cyber doctrine can only be realized "when commanders consider and employ capabilities across" all three levels (Electronic Warfare 2007).

METHODOLOGY

In order to infer China's cyber warfare doctrine, it is necessary to examine the complete history of Chinese developments relating to cyber warfare. This research analyses all pertinent open source material and identifies capabilities, goals, tactics, training, trends, and strategy. These are compared with China's culture, state-level goals, and existing broad defense doctrine, including how it has adapted and transformed itself to date. A three-branch model of cyber warfare is employed to organize overlapping terminology and concepts, while adhering to China's all-inclusive approach. This book identifies and maps the major Chinese government and military organizations responsible for cyber warfare activities. A complete list of alleged computer network exploitation incidents is provided and analyzed, including an assessment of the validity of these claims. Additionally, the three branches of cyber warfare are applied to the Taiwan issue, in order to demonstrate how China would use these in combination to achieve its objectives.

DEFINITIONS

Central to this book is the integration of a plethora of overlapping and vague terminology. Portions of this section have been adapted from the 2013 East Asia Security Symposium and Conference paper titled *The Semantics of Cyber Warfare* and improved upon for this book (see Fritz 2013b).

With the growth in cyber warfare as a field of study, the amount of cyber-related terminology has also grown. The use of this terminology in prominent Chinese, and Western, literature is riddled with contradictions. Researchers do not consistently agree with themselves, let alone others. It is common to find authors who proclaim a difference between specific terms, yet they then proceed to use them interchangeably. Others will introduce new terms which are unnecessary in view of viable existing terms. Reasons for such discrepancies include disagreements among translations, PLA writings inconsistently incorporating or referencing U.S. doctrinal writings, and the relative youth of this discipline (Mulvenon 1999; Thomas 2004, 3). Both China and the United States have multiple government and military agencies putting forward their own definitions and competing to take the lead in this field within their respective states. As a further complication, China lacks transparency of endorsed cyber doctrines. This makes contemplating one definitive set of definitions problematic, as there are a multitude of them even within a single military branch.

Shen Weiguang, "the father of Chinese IW [Information Warfare]," criticized the United States for not having a universal definition, yet he then proposed vague terms such as information alliances, information arena, information borders, information deterrence, information factories, information fighting, information firepower surveillance, information frontier, information invasion, information pollution, information resources, information space, information territory, information weapons, information war, and separately, war in the information age (Thomas 2004, 32, 35–36, 38–40, 42, 51). Major General Dai Qingmin, former head of the General Staff Department's Fourth Department, authored the 2002 book *Direct Information Warfare*. In it he introduced the terms battlefield information environment, computer network space (or bit battlefield), digitized armed forces, electronic camouflage, electronic/intellectual/thought-information warfare, information colonialism, information mobilization, information supremacy, network psychological warfare, network warfare, people's network war, and virtual warfare (Thomas 2007, 117–40). Adding further to the confusion are the introduction of terms by other prominent Chinese authors such as chip warfare, informatized warfare, intangible war, masquerade technology, media warfare (including morale bombs), system attack warfare, take home battle, and technology aircraft carriers (Thomas 2004, 2, 60, 65; Thomas 2007, 87, 91,

244; Thomas 2009, 38, 48). All of these have overlapping, unclear, contradictory, and/or missing elements, and their exact definition varies between authors. As another example, some PLA officers distinguish between cyberspace, digital space, and information space (Wortzel 2014). Clarity deteriorates further as aspiring theorists attempt to integrate this tapestry of terms into their own work. Even the most well established and frequently used terms are found jumbled together without clear distinction. For example, in the span of one short entry, The *Military Balance 2015* describes China's cyber developments by using the terms computer network operations (and its components), cyber warfare, electromagnetic forces, electronic warfare, information confrontation (xìnxī duìkàng; 信息对抗), information operations, information warfare, informationized conditions, and integrated network electronic warfare (INEW—wǎng diàn yì tǐ zhàn;一网电一体战) (International Institute for Strategic Studies 2015, 246; see also Krekel, Adams, and Bakos 2012; U.S. Department of Defense 2010). While the authors identified in this introductory chapter are undoubtedly respected leaders in this field, and their writings are exceptional on many levels, this book requires the ability to integrate the full range of cyber warfare literature and cannot adhere to the multitude of each author's individual preference for definitions. Instead, in the following conceptual endeavor, they are merged on the basis of core distinctions and majority trends.

This book has organized all cyber warfare[3] terminology into three fundamental branches that are common to both the People's Republic of China and the United States of America. Rather than introduce new terms, existing "best-fit" terms are used. These are: Computer Network Operations (CNO), Information Operations (IO), and Net-Centric Warfare (NCW). Although these terms are U.S.-centric, they carry the benefit of having been used extensively in authoritative open source material over a relatively long period of time. Regardless of which name is attached to these branches, they are structurally distinct, and all previous terms can be placed within one of these three, thereby helping to bring clarity to this field. While the terms chosen may be open to debate, the category they denote below remains valid. As shown in figure 0.1 and table 0.1, their distinguishing characteristics are whether or not they are connected to the Internet, involve hacking, or involve military hardware. These may appear to be obvious distinctions, yet they remain absent from prominent definitions. It is important to note that the terms cyber warfare and Information Warfare (IW) will be considered interchangeable, and they represent the broadest term encapsulating the other three.

All three branches can be used in combination, as well as in combination with physical acts, yet their differences and the scope of this field warrant classification. CNO includes computer network attack (CNA), computer network exploitation (CNE), and computer network defense (CND). The defining characteristics of CNO are hacking (unauthorized

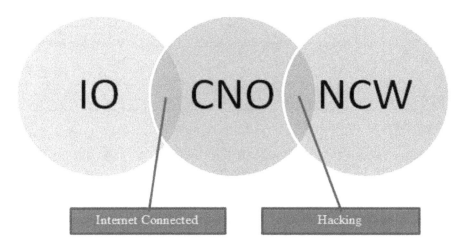

Figure 0.1. Linear Venn of Cyber Warfare Branches. *Fritz 2013b.*

access) into a computer or network, via the Internet. If these two criteria are met, the remainder of the definition is fairly consistent among authors; although, some may also use additional terms interchangeably. CNA includes the ability to deceive, degrade, deny, destroy, or disrupt computers and networks through common hacker methods like denial of service, Trojans, viruses, worms, and so forth (these are examined in detail in chapter 1). CNE is typically theft, eavesdropping, and setting up for an attack, while CND is defending against attack or exploitation. Information Operations entails utilizing and manipulating information online, without hacking, to influence actions. Examples of IO include online censorship, propaganda, psychological operations (PSYOPS), and recruitment. This categorization fits closely with a declassified 2003 document by the U.S. Department of Defense, titled *Information Operations Roadmap*. The *Information Operations Roadmap* includes CNO and Electronic Warfare by name. However, these two elements have continued to develop in their own right, and the novel portion of IO within the *Information Operations Roadmap* was the Internet PSYOPS and non-hacking operations. Later publications on IO doctrine endorsed by the U.S. government and U.S. military have continued to alter its overall framework, while maintaining a clear connection to the more distinct and stable definition used in this book (see Information Operations 2012; Military Information Support Operations 2010). NCW refers to advanced military weapons systems, communication, and situational awareness that utilize closed computer networks (networks which are not connected to the Internet). It concerns the application of information and communication technology to enhance combat effectiveness and efficiency. These

Table 0.1. Comparison of Cyber Warfare Branches

	Computer Networks	Internet Connected	Hacking	Military Hardware
CNO	Yes	Yes	Yes	No
IO	Yes	Yes	No	No
NCW	Yes	No	Yes	Yes

Updated from Fritz 2013b

computer networks can be hacked into, but not through the Internet. Additionally, NCW encompasses electronic warfare.

How do these three branches fit in with the most commonly cited Chinese terms? Key terms prevalent in Chinese government and military literature include information confrontation, information operations, information warfare, informationization, integrated network electronic warfare, and system of systems operations (tǐxì zuòzhàn; 体系作战). In general, Chinese military theorists use information operations and information warfare as catchall terms equivalent to cyber warfare. This book concurs with information warfare being a synonym of cyber warfare; however, it differs by using the stricter definition of IO given above. Chinese literature has many terms which fit within that definition, yet none which can replace IO as the branch name in terms of breadth, precision, and unanimity. Informationization is equivalent to NCW. A large body of evidence for this conclusion, and justification for using NCW in its place, is given in chapter 4. Additionally, China's system of systems operations is directly taken from prior NCW research. The two remaining Chinese terms which are most frequently referenced in Chinese cyber warfare literature are integrated network electronic warfare and information confrontation. INEW is a combination of computer network attack and electronic warfare, which makes it a combination of subcomponents from within the branches of CNO and NCW. Justification for INEW not being used as a primary branch of cyber warfare will be provided in chapter 4. Among the reasons given are that the PLA appears to be addressing some of INEW's drawbacks by moving toward the concept of information confrontation. Information confrontation has been described as using CNO, IO, and NCW in combination,[4] which makes it synonymous with cyber warfare. Further, the Chinese characters which are translated as "information confrontation" in some English documents are also translated elsewhere as "information warfare," which has already been labelled above as a synonym of cyber warfare. The PLA is continuing to develop and refine these concepts, so this term might not last either. For these reasons, this book prefers to use the pre-existing term cyber warfare. What the name represents is more important than the name itself, and these decisions are applied consistently throughout this book. (See figure 0.2 for an overview of the synonyms addressed above and the location of these terms in relation to each other.)

DELIMITATIONS AND LIMITATIONS

This book has five primary delimitations and one limitation. First, this research will not delve into the pre-existing debate over the definitions of "hacking" and "cyber war" as these are discussed at length by other authors and these words have already entered common use by officials and the media. Second, all Internet-connected computers and networks have the potential to be attacked, and an intruder can gain up to complete control of these systems. The number of potential targets, types of attack, and tactics are therefore enormous, and a complete list of theoretical possibilities is beyond the scope of this book. Chapter 1 will seek to identify general trends, pivotal examples, and innovative applications of CNA. Unlike CNE, however, it does not have a catalogue of alleged incidents as a guide. As of 2016, few high-level state-sponsored attacks have occurred (reasons for this are provided in chapter 1). Third, this work's research into IO is focused on its application toward foreign audiences, as opposed to domestic audiences, such as domestic censorship. There are some important interconnections where one impacts on the other, and these are addressed accordingly. IO also crosses into other areas of study, such as advertising, marketing, and psychology; neverthe-

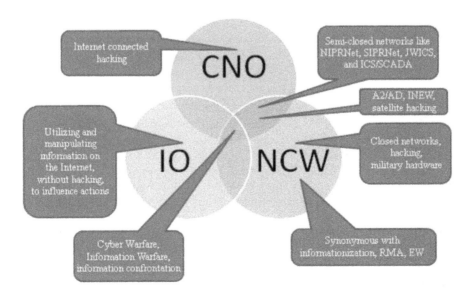

Figure 0.2.   Venn of Cyber Warfare with Select Callouts. *Updated from Fritz 2013b.*

less, these surpass the extent of this research. Fourth, NCW deals with military computers and networks, and attempts to link these systems together, or alternatively, attempts to hack into these systems. This includes all command and control, communication, sensor, and weapon systems, across all domains of warfare in all branches of the armed forces. As a result, it encompasses an immense diversity of hardware and software technical specifications. This book will focus on central concepts and strategy, rather than technical specifications. Fifth, this wide range of computer network architectures, coupled with the details of many being designated as classified information, make it difficult to ascertain which networks share a connection with the Internet. Evidence suggests that any battlefield networks which do share an Internet connection are in the process of being separated. Finally, this research is limited by the availability of Chinese-language open source material. Exhaustive efforts and consideration were given to the identification and translation of Mandarin language source material, yet researchers who are fluent in both English and Mandarin may discover valuable information not present here. However, this seems unlikely given this research's focus on prominent authors within the cyber warfare community, as well as globalization and the Internet allowing information to spread, and governments and militaries setting a boundary to available information through the demarcation of classified material. Despite these delimitations and limitation, this book addresses a serious gap in both cyber warfare studies and China's strategic doctrine. China's momentous emergence into the international sphere has been subject to intense scrutiny. However, much needs to be done to understand the strategic context, the conceptual framework and historical evolution that have shaped its development of cyber warfare doctrine.

OUTLINE OF THE BOOK

This book is divided into five chapters. Chapter 1 examines the development of Chinese computer network operations, including the structure of government entities responsible. Chinese theoretical work on CNO began in the mid to late 1980s. The creation of the PLA's first CNO unit took place as early as 2000, and as of 2016, the PLA is believed to possess at least 20 units. China is also believed to possess approximately 33 CNO militia and 250 hacker groups with their formation having begun in the late 1990s. Computer network attack and exploitation can enhance China's ability to conduct asymmetric warfare and leapfrog in modernization. At the same time, China is becoming increasingly reliant on Internet-connected computer networks, and attacks against China are rapidly increasing. Chapter 2 traces the foundation of Chinese hacker groups at the turn of the century. These groups set the stage for Chinese computer network exploitation, which became dominant around 2002. This chapter will explore over 100 isolated CNE incidents and 19 advanced persistent threats, targeting 71 countries, all of which are allegedly attributed to China. CNE can provide substantial economic gains at a low cost, although, international backlash has risen since 2010. Chapter 3 details China's development of Information Operations. These non-hacking Internet activities can influence actions in an attempt to attain "victory without war" (Thomas 2004, 37). IO uses censorship, media, propaganda, and PSYOPs to shape opinions. These suit China's holistic approach to cyber warfare and Sun Zi's views on deception, by attempting to control conduct, rather than resorting to the use of direct force. This chapter also examines which government entities are responsible for overseeing Chinese IO. Chapter 4 investigates Chinese Net-Centric Warfare. Central to this is establishing NCW's connection to the synonymous terms of Revolution in Military Affairs and informationization, and demonstrating that NCW encompasses electronic warfare and anti-access area denial. Each of these enhances understanding of the development of Chinese NCW. The application of ICT to military operations is linking together command, equipment, and soldiers on the battlefield, under a system of systems, resulting in increased combat effectiveness and efficiency. However, these non-Internet-connected computer networks are susceptible to hacking. NCW is concerned not only with advances in technology, but also changes in doctrine, training, and organization tailored to make optimal use of these technologies. Chapter 5 examines the application of the three cyber warfare branches to the Taiwan issue. It seeks to demonstrate how these branches are distinct, yet interconnected, and how they create synergy, which can aid China in achieving its goals. Lastly, the conclusion covers this book's findings and states their implications.

NOTES

1. The 2010 *Department of Defense Dictionary of Military and Associated Terms* defines branch as "a subdivision of any organization," or relative to the ground warfare domain, as "an arm or service of the Army."

2. In 2014, the PLA possessed one aircraft carrier, the *Liaoning*, which was commissioned in 2012.

3. Cyber warfare is the use of cyberspace to engage in war or conflict, or to engage in activities relating to war or conflict, including altering the balance of power or enhancing defense. The 2013 U.S. doctrinal publication, *Cyberspace Operations*, defines cyberspace as "the interdependent networks of information technology infrastructures and resident data, including the Internet, telecommunications networks, computer systems, and embedded processors and controllers," and "the content that flows across and through these components." It is important to note that this includes, but is not limited to, the Internet (see Andress and Winterfeld 2011, 2–5; *Cyberspace Operations* 2013; Department of Defense Dictionary of Military and Associated Terms 2010).

4. These branch names are not always used directly; terms which this book has deemed to be their equivalent or subcomponents are. For example, "network warfare [NCW], electronic warfare [a subcomponent of NCW], psychological warfare [IO,

based on the context]", and "cyber warfare [CNO, based on the context]" are named as elements of information confrontation (Krekel, Adams, and Bakos 2012; Wortzel 2014). The U.S. Department of Defense's 2014 Annual Report to Congress provides another example of inferring meaning. It states that "China's investments in advanced electronic warfare (EW) systems [NCW], counterspace weapons, and computer network operations (CNO)—combined with propaganda [IO] and denial through opacity— reflect the emphasis and priority China's leaders place on building capability for information advantage" (U.S. Department of Defense 2014).

List of Abbreviations

3PLA	GSD Third Department
4PLA	GSD Fourth Department
A2/AD	Anti-Access Area Denial
ADIZ	Air Defense Identification Zone
AEW&C	Airborne Early Warning and Control
AIT	American Institute in Taiwan
AMS	Academy of Military Science
AMSC	American Superconductor
APEC	Asia-Pacific Economic Cooperation
APT	Advanced Persistent Threat
ASAT	Anti-Satellite weapon
ASB	AirSea Battle
ASBM	Anti-Ship Ballistic Missile
ASCM	Anti-Ship Cruise Missile
ASEAN	Association of Southeast Asian Nations
ASIO	Australian Security Intelligence Organisation
BfV	(German) Federal Office for the Protection of the Constitution
C2	Command and Control
C4ISR	Command, Control, Communications, Computers, Intelligence, Surveillance and Reconnaissance
CAC	Cyberspace Administration of China
CIS	Commonwealth of Independent States
CMC	Central Military Commission
CNA	Computer Network Attack
CND	Computer Network Defense
CNE	Computer Network Exploitation

CNERT/CC	National Computer Network Emergency Response Technical Team / Coordination Center of China
CNO	Computer Network Operations
CNP	Comprehensive National Power
CPC	Communist Party of China
CPD	Central Publicity Department, formerly Propaganda Department
CSSTA	Cross-Strait Service Trade Agreement
DDoS	Distributed Denial-of-Service
DRDC	Defence Research and Development Canada
DSB	Defense Science Board
DOI	Department of Informationization, formerly Communications Department
EA	Electronic Attack
ED	Electronic Defense
EADS	European Aeronautic Defence and Space company
EME	Electromagnetic Environment
EMS	Electromagnetic Spectrum
ES	Electronic Surveillance (or Electronic Warfare Support)
ETIM	East Turkistan Islamic Movement
EW	Electronic Warfare
FBI	Federal Bureau of Investigation
FBIS	Foreign Broadcast Information System
FCS	Future Combat Systems
GAD	General Armaments Department
GIG	Global Information Grid
GLD	General Logistics Department
GPD	General Political Department
GPS	Global Positioning System
GSD	General Staff Department
HAARP	High Frequency Active Auroral Research Program
HALE	High-Altitude Long Endurance
HUMINT	Human Intelligence

IAB(a)	Information Assurance Base
IAB(b)	Internet Affairs Bureau
IADS	Integrated Air Defense System
ICS	Industrial Control Systems
ICT	Information and Communication Technologies
IT	Information Technology
IMF	International Monetary Fund
INEW	Integrated Network Electronic Warfare
IO	Information Operations
IOD	Information Operations Defense
IOE	Information Operations Exploitation
IP	Internet Protocol
IRC	Internet Relay Chat
ISP	Internet Service Provider
ISR	Intelligence, Surveillance and Reconnaissance
IW	Information Warfare
IWM	Information Warfare Militia
JWICS	Joint Worldwide Intelligence Communications System
MFA	Ministry of Foreign Affairs
MIIT	Ministry of Industry and Information Technology
MIRV	Multiple Independently Targetable Re-entry Vehicle
MND	Ministry of National Defense
MOC	Ministry of Culture
MOE	Ministry of Education
MOFCOM	Ministry of Commerce
MOOTW	Military Operations Other Than War
MPS	Ministry of Public Security
MR	Military Region
MSS	Ministry of State Security
NAPSS	National Administration for the Protection of State Secrets
NASA	National Aeronautics and Space Administration
NATO	North Atlantic Treaty Organization

NCPH	Network Crack Program Hacker
NCW	Network-Centric Warfare
NDU	National Defense University
NGO	Non-Government Organization
NIPRNet	Nonsecure (or Non-classified) Internet Protocol Router Network
NSA	National Security Agency
OFP	Office for Foreign Propaganda
OSINT	Open Source Intelligence
PLA	People's Liberation Army
PLAAF	PLA Air Force
PLAGF	PLA Ground Force
PLAN	PLA Navy
POS	Point of Sale
PRC	People's Republic of China
PSYOPS	Psychological Operations
R&D	Research and Development
RFA	Radio Free Asia
RFID	Radio-Frequency Identification
RMA	Revolution in Military Affairs
SAARC	South Asian Association for Regional Cooperation
SAC	Second Artillery Corps
SAPPRFT	State Administration of Press, Publication, Radio, Film and Television
SC	State Council
SCADA	Supervisory Control And Data Acquisition
SCIO	State Council Information Office or Information Office of the State Council
SCO	Shanghai Cooperation Organization
SIGINT	Signals Intelligence
SIIO	State Internet Information Office
SIPRNet	Secret Internet Protocol Router Network
SLOC	Sea Lines Of Communications
SOE	State-Owned Enterprise

SRBM	Short-Range Ballistic Missile
TT&C	Tracking, Telemetry, and Control
UAV	Unmanned Aerial Vehicle
UCAV	Unmanned Combat Aerial Vehicle
UGV	Unmanned Ground Vehicles
USCC	U.S.-China Economic and Security Review Commission
USNSS	United States National Security Strategy
USTRANSCOM	United States Transportation Command
VOA	Voice of America
VoIP	Voice over Internet Protocol
VSAT	Very Small Aperture Terminal

ONE

Computer Network Operations

Computer network operations (CNO) includes computer network attack, computer network exploitation, and computer network defense. The two defining characteristics of CNO are (1) hacking into a computer or network, (2) via the Internet. Computer network attack (CNA) includes attempts to disrupt, deny, degrade, deceive, or destroy; and it utilizes traditional hacker tools like viruses and worms.[1] Examples of CNA include causing a computer to become inoperable, deleting or re-writing of files, and knocking a network or service offline. Computer network exploitation (CNE), also known as cyber espionage or cyber reconnaissance, encompasses the theft of intellectual property and intelligence gathering. CNE also includes mapping and identifying weaknesses in opponent networks, as well as establishing a covert presence in these networks, all of which can be used to aid CNA. In many cases an intruder can go from exploitation to attack at will, as they utilize many of the same tools and skillsets. Consequently, in addition to the outcome of an operation, the perceived intent of an intruder can be used to distinguish between a CNA and CNE incident. CNE has been far more prevalent than CNA among alleged state-sponsored CNO. It provides significant benefits, particularly for China the transfer of intellectual property, with minimal consequence in comparison to CNA. Computer network defense (CND) is defending against attack or exploitation, including attempting to identify intruders, sometimes referred to as digital forensics. CND debatably fits the criteria for Information Operations (IO), "using the Internet to one's advantage without the use of hacking" (see chapter 3); however, CND is more firmly rooted in CNO and is widely considered to be a component of CNO. Conducting effective defense requires a deep understanding of attack and exploitation, and vice versa, therefore mastering any one of the three elements of CNO will yield knowledge on how to conduct the other two.

Beyond standard Internet-enabled networks, there are some variants which can be hacked into and are considered a part of CNO, yet they challenge the boundaries of CNO's definition. These include quasi-closed networks, networks which are partially, but not entirely, separated from the Internet. For example, military and intelligence closed networks such as the United States' Nonsecure Internet Protocol Router Network (NIPR-Net), Secret Internet Protocol Router Network (SIPRNet), and Joint Worldwide Intelligence Communications System (JWICS). The term closed network, also known as a network air gap,[2] is applied loosely here as open source material is unclear on exactly how separate these networks are from the Internet. These are highly secure networks; however, they appear to be remotely accessible online (either directly or by intermediary connections), use much of the same infrastructure as the Internet, and are vulnerable to traditional hacking tools. For example, U.S. military forces have reportedly used civilian infrastructure to connect to NIPRNet and then used tunnelling protocol[3] to connect from NIPRNet to SIPRNet (Wilson 2007). Further, while these networks aid in military operations, they serve a much larger role than being a part of a weapon system. China has also pursued research and development of a closed, or China-only, Internet for classified information, although its progress is unknown (Thomas 2004, 10–11).

Industrial Control Systems (ICS), including Programmable Logic Controllers (PLCs) and Supervisory Control and Data Acquisition (SCADA) systems, are also quasi-closed networks that fit most appropriately in the cyber warfare branch of CNO. These computer networks are commonly associated with services such as electricity distribution, nuclear power, oil and gas pipelines, waste management, and water treatment. The technical specifications for ICS hacking are different than that of traditional hacking, however it shares much more in common with CNO than it does with the military hardware of Net-Centric Warfare (NCW). The Stuxnet worm discovered in 2010, which targeted uranium enrichment facilities in Iran, illustrates the incomplete isolation of ICS networks. USB drives containing the Stuxnet worm were inserted into computers belonging to the allegedly closed networks. After the networks became infected, the attackers were able to update remotely and exfiltrate data using computers located in foreign countries, computers which were not a part of the original closed networks, suggesting a capability of Internet connectivity prior to the infiltration. A truly closed network would have no wireless transmission and share no physical connections to Internet infrastructure. Further, while some of these networks are at least semi-closed off from the Internet, many are not and do not claim to be. Beyond cyber warfare, most states consider ICS and SCADA to be critical infrastructure. Other critical infrastructure such as banking, transport, medical facilities, emergency services, and civilian communication are more firmly rooted in the Internet, and therefore fall under the CNO category when hacked. For these reasons, it is appropriate and efficient to keep ICS together with the rest of critical infrastructure in cyber warfare categorization. Other specialized networks with distinct configurations, hardware, or means of transmission, such as mobile phones, Point of Sale (POS) systems, and satellites, also fall under the CNO category when hacked and will be considered in this chapter.

CNO suits multiple pre-existing Chinese doctrinal characteristics, and it provides advantageous means for advancing China's goals. One of these characteristics epitomized by CNO is the PLA's emphasis on asymmetric warfare. For example, China's perceived top competitor and military superior, the United States, relies heavily on computer networks for government and military operations, and has a commercial sector and civilian population that are reliant on an unsecure computer infrastructure. CNA and CNE have the benefits of plausible deniability, great impact at a relatively low cost, a low entry level,[4] high speed, the near removal of geographic distance and borders, stealth, and an underdeveloped international legal framework to combat against it. The requirements for successful CNA are much lower than the requirements for successful CND. A defender must identify and protect against all possible vulnerabilities, while an attacker needs to only find one vulnerability to cause damage. Meanwhile, CNE follows Beijing's predilection toward technology transfer and reverse engineering as a means of leapfrogging in modernization and international competitiveness (see Fritz 2008). Online industrial espionage allows China the opportunity to skip generations of research and development efforts, reaching comparable levels in science and technology, and by association boosting economic and military might. Economic gains strengthen China's overall condition while in relative terms weakening that of its competitors, and it helps to ensure the stability and legitimacy of the ruling Communist party. Cataloguing adversary weaknesses through online reconnaissance not only provides an asymmetric advantage in the event of a conflict, the threat it poses also acts as a deterrent and distraction while China catches up in traditional military might. Further, CNE can aid China in obtaining market dominance within the fields of Information and Communication Technology (ICT). This would provide increased CND by removing foreign influence, and enhance CNA capability, such as allowing China to pre-install exploits[5] in ICT products or expand ownership of Internet infrastructure. Market dominance also relates to financial gain, which China has stated is intrinsically related to military capabilities and strategic interests. China's view of comprehensive national power, and concepts such as unrestricted warfare (discussed in the introduction), has shown a blurring of the lines between military and non-military spheres. These correlate with CNO, as anyone connected to the Internet can now participate in a state-level conflict. Additionally, non-military computer networks have become appealing targets, because they are vital to a state's ability to function, and their disruption or destruction would not directly involve harming civilians. Additional Chinese strategic inclinations which can work in harmony with CNO will be explored throughout this chapter's subheadings of: organizational composition, computer network attack, computer network exploitation, and computer network defense. Lastly, China's attempts to influence the development of CNO-related international law will also be explored.

CNO ORGANIZATIONAL COMPOSITION

In 2011, China's Ministry of National Defense announced the creation of an "online blue army," saying that it was in an early development stage, primarily defensive, and in response to the threat posed by other states' development of CNO units (Guo, Gu, and Wu 2011). The announcement garnered international attention; however, multiple PLA-endorsed publications place the formation of the first Chinese state-sponsored CNO units approximately a decade prior, between the years 2000 to 2003. While the terminology used in some of these publications could be referring to NCW units rather than CNO, a report presented by PLA representatives at the 10th National People's Congress in 2003 specifically referred to the ability to conduct cyber warfare "on the Internet" (Thomas 2004, 61). Chapter 2 of this book, *The Development of Computer Network Exploitation, 1998–2014*, also details over 100 alleged state-sponsored incidents dating back to 1999. The seemingly overdue announcement of an online blue army could be propaganda, or a slight toward allegations of Chinese CNE, yet it is significant in establishing China's publicly endorsed baseline for the development of CNO. The online blue army was comprised of 30 operators, based in Guangdong province, and possessed a budget of approximately 1.5 million U.S. dollars (Beech 2011; Lewis 2011).[6] Further, China's 2013 edition of *The Science of Military Strategy* acknowledges the existence of defensive and offensive CNO units within the PLA (Gady 2015; Tiezzi 2015). Western government agencies, Internet security firms, and think tanks place the number of Chinese state-sponsored CNO units at approximately 20, spread across all seven PLA Military Regions (MRs), and with personnel ranging from hundreds to thousands per unit (Clayton 2012; Mandiant 2013; Riley and Dune 2012; Segal 2012). Moreover, these figures are restricted to PLA units and do not include other government agencies which may be conducting CNO, hacker groups with government ties, militias, or the readiness of China's general population to conduct online "people's war" (discussed in the introduction of this book).

A wide range of Chinese entities appear to play a role in state-sponsored CNO. For example, the Ministry of Public Security and the Ministry of State Security have been "authorized by the military to carry out network warfare operations," and a group within the Ministry of Industry and Information Technology was "tasked with breaking computer codes of foreign companies and governments" (Gady 2015; Henderson 2007a; Tiezzi 2015). However, the General Staff Department's (GSD's) Third and Fourth departments are allegedly China's primary organs for conducting CNO (International Institute for Strategic Studies 2015; Krekel, Adams, and Bakos 2012; Mandiant 2013; Stokes, Lin, and Hsiao 2011). The Fourth Department (4PLA), which has traditionally dealt with electronic warfare, is allegedly responsible for CNA, while the Third Department (3PLA), which has traditionally dealt with signals intelligence, is allegedly responsible for CNE and CND (see figure 1.1). While these departments appear to play a primary role in CNO, they are not the only actors, and it might represent a continually developing attempt to centralize authority over China's cyber warfare capability. Of these two departments, the 3PLA reportedly delegates operations to its technical reconnaissance bureaus (TRB) spread across the mainland (Mandiant 2013; USCC Annual Report 2009). Each of China's four armed service branches—the PLA Ground Force, the PLA Navy (PLAN), the PLA Air Force (PLAAF), and the Second Artillery Corps—along with the People's Armed Police (PAP), also appear to conduct some level of in-house CNO activity to support their specific needs. Further, their research and development, including war games, have an impact on overall Chinese CNO doctrine. Additionally, the Second Department of the GSD, which is responsible for military intelligence, and key government agencies such as the Ministry of State Security and the Ministry of Public Security, would benefit from a CNO capability and appear to at least act in a supporting role, such as providing input on target selection (USCC Annual Report 2012). This complex and changing structure for coordination is comparable to the United States' difficulties in establishing a leading organization for CNO amid multiple competing actors.[7] Streamlining computer network operations was among President Xi Jinping's agenda when he presided over the first meeting of China's Internet security and infomationization[8] leading group in 2014 (China's New Small Leading Group on Cybersecurity and Internet Management 2014; Xi Jinping leads Internet security group 2014). Further, President Xi specifically advocated "taking inspiration" from similar attempts at improving CNO organizational structure taking place within the United States and Japan (Panda 2014).

Militias and hacker groups also factor into Chinese CNO planning. According to articles published by the *Liberation Army Daily*, Chinese cyber warfare theorists advocated the formation of militia proficient in CNO as early as 1997 (Thomas 2004, 7, 9). *China National Defense News* reported that CNO militia became a reality in 1999 and 2000, with units

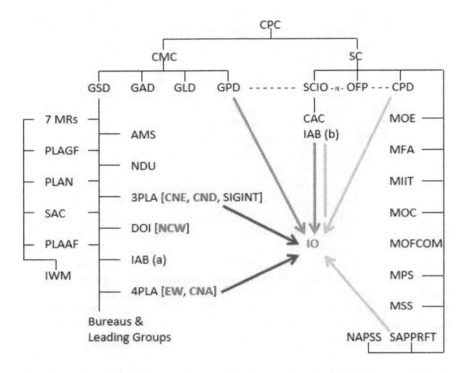

Figure 1.1.   Key Government Entities Responsible for the Development of Chinese Cyber Warfare. The SCIO and OFP operate under a dual nameplate system. Dashed lines represent close ties/brother organizations. This chart is not comprehensive; it only notes key departments which are recurringly referenced as being relevant to cyber warfare. MIIT, MOC, MPS, and MSS have been named as participating in some aspects of CNO, and the armed service branches will inevitably have some input in the development of NCW; however, the primary departments responsible for those cyber warfare branches are currently the 3PLA and 4PLA. IO has a wide range of participants whose inclusion varies by source. The most probable collection of IO actors is the GPD and SCIO. The CAC, formerly SIIO, is a dual nameplate with the Office of the Central Leading Group for Cyberspace Affairs, formerly the Internet Security and Informationization Leading Group. CNO (Gady 2015; Henderson 2007a; International Institute for Strategic Studies 2015; Kelly and Lee 2012; Krekel 2009; Krekel, Adams, and Bakos 2012; Mandiant 2013; Novetta 2014; Stokes and Hsiao 2012; Tiezzi 2015; USCC Annual Report 2012). NCW (Gady 2015; Information Office of the State Council of the People's Republic of China 2013; Krekel 2009; Krekel, Adams, and Bakos 2012; Mandiant 2013; Tiezzi 2015). IO (GPD and SCIO) (Fritz 2008; HBGary 2011; Wortzel 2013). IO (SC) (Chang 2015; Cheung 2015; China orders media giant Sina to "improve censorship" 2015; China's New Small Leading Group on Cybersecurity and Internet Management 2014; China's Propaganda and Influence Operations 2009; Chinese websites to "spread positive energy" 2013; Creemers 2014; Creemers 2015; Cyberspace Administration of China 2015; Cyberspace Administration of China launches official website 2015; Feakin 2013; Glanz and Markoff 2010; Hansen 2014; Jing 2014; Kelly and Lee 2012; Lam 2013a; Lawrence and Martin 2013; Panda 2014; Rawlinson 2015; Rumi 2004; Segal 2014; Shambaugh 2007; Tham 2015; The Organizational Structure of the State Council 2003; Wines 2011; Xi Jinping leads Internet security group 2014). IO (GSD) (China's New Small Leading Group on Cybersecurity and Internet Management 2014; Feakin 2013; Krekel, Adams, and Bakos 2012)

operating in Datong, Xiamen, Xian, and the Echeng District (Thomas 2004, 9–10). Personnel are pooled from the information technology industry and academia, and they maintain their separate careers while acting as a type of reserve force (Thomas 2007, 129). In 2005, the PLA General Political Department reported the creation of 290 cyber militia in the Chengdu military region alone (Krekel, Adams, and Bakos 2012). However, this figure did not delineate between CNO, NCW, and IO militia, of which the monitoring, censorship, and propaganda of IO would seem to require the most people and the least technical expertise. A 2008 study by Verisgn's iDefense security intelligence services estimated that there were 33 CNO-specific cyber militia in operation countrywide (USCC Annual Report 2009). In addition to these militia aiding China's CNO capability and readiness, they also appear to actively engage in CNE. China's 2013 National Defense White Paper stated that China's armed forces seek to enhance "integrated civilian-military development" and "reserve force building" (Information Office 2013). China also possesses an estimated 250 hacker groups, who can develop cyber weapons and advance methods of CNO, serve as a recruitment pool, provide a supporting role in operations, or act on behalf of the government while providing plausible deniability (China's Proliferation Practices, and the Development of Its Cyber and Space Warfare Capabilities 2008; USCC Annual Report 2008; USCC Annual Report 2009).[9]

Additional collaboration by entities, individuals, and the general population as a whole further complicate the organizational structure of China's CNO capability. Key institutes, including those directly endorsed by the PLA, such as the Academy of Military Sciences and the National Defense University, as well as civilian universities with special military associations and benefits, promote the advancement of CNO research and the training of potential recruits. Government-funding mechanisms, including the 863 and 973 programs, also foster the development of CNO-relevant research, such as encryption and "data mining techniques" (USCC Annual Report 2012). Independent Chinese students and business people with access to key individuals or networks abroad, and entities like border control (customs) and telecommunications firms, may be enlisted on a temporary basis to enhance specific CNO operations (USCC Annual Report 2012; USCC Annual Report 2014). Chinese companies, such as state-owned enterprises, might even take it upon themselves to conduct online industrial espionage, in order to meet government expectations and improve their standing with government officials. Lastly, China possesses the world's largest online population and could use this to its advantage in online people's war. Unlike prior iterations of people's war, CNO provides China with the ability to take action globally, not just in physical defense of the homeland. It also diminishes the risk of physical harm associated with traditional warfare, thereby reducing hesitation to participate. Further, it suits China's overall nationalistic sentiment, favors the urban (more connected) population, and does not necessarily require specialized computer skills. Group leaders could disseminate software online which helps automate the attack process, including target lists and malicious scripts. In this way, China would be relying more on the people's computing power than their technical expertise.

COMPUTER NETWORK ATTACK

An online intruder can obtain full control over a victim's computer, so at its maximum computer network attack (CNA) is capable of any harmful action that a user could do to their own computer via the keyboard. Despite great concern over CNA displayed by governments and the media, there are few allegations of state-sponsored CNA in open source material, particularly in comparison to the large quantity of alleged state-sponsored computer network exploitation. Even in the case of the 2007 cyber attacks against Estonia and the 2008 cyber attacks surrounding the Russo-Georgian War, for which Russia was accused, the assailants did not cause the dramatic destruction of computer networks. Instead the perpetrators relied on low-level attacks such as DDoS[10] and web page defacements.[11] This could be due to states not wanting to reveal their capability until it is needed, such as in a time of war with a superior or peer-level adversary; or it could be the result of an attack carrying greater perceived consequences than exploitation. For example, using CNA could escalate a cyber arms race; or the aggressor state may be heavily reliant on computer networks themselves and fear setting a precedent that would reduce the hesitation of other states to conduct CNA against them. Additionally, the benefits of conducting CNA may not be worth the risk of hindering continued intelligence gathering and intellectual property theft obtained through online exploitation. In this sense CNA could be comparable to anti-satellite (ASAT) weapons. The United States and Russia (formerly the Soviet Union) have had ASAT capability since the 1970s, and China has had it since 2007; however, none of these states have deployed them in combat. For the United States and the Soviet Union during the Cold War, using an ASAT could have escalated to nuclear war, and it would put at risk all of the benefits obtained from properly functioning satellites, including the intelligence gained from satellite imagery. Despite limited international laws preventing the use of ASATs, as is the case with cyber weapons, the perceived blowback acted as a sufficient deterrent. This does not mean, however, that the cost-benefit analysis could not change during heightened circumstances or that states would not continue to develop and maintain an active capability.

The 2013 National Defense White Paper states that China's policies of active defense, "not attacking unless attacked," and a resolute determination and capability to "counterattack if attacked," apply to cyberspace (Information Office 2013). While on the surface these appear defensive in nature, the determination of terminology—such as an act of aggression, imminent threat, threat to national security, or state secrets—remains ambiguous, and could lead to actions that would be viewed by other states as an act of war or a pre-emptive strike. Some PLA-affiliated cyber warfare theorists and Western analysts believe that China's initial targets in a cyber war would be the "critical nodes and control centers" which link the "political, economic, and military installations of a country (as well as society in general)" (Thomas 2004, 6–7). These nodes are also responsible for enabling a state to make effective decisions and coordinate actions. China believes the United States set a precedent for such an attack in the 1991 Gulf War by allegedly using online computer viruses to destroy Iraq's air defense system; although Western states hold a different account of this event (Guo, Gu, and Wu 2011; Mulvenon 1999). Critical nodes which could be struck are not restricted to military infrastructure and could include vital financial or societal networks which would have a debilitating effect. Chinese cyber warfare literature often describes this as "acupuncture war," "key-point strikes," or targeting an information-reliant state's "Achilles' heel" (Andress and Winterfeld 2011, 43; Cliff 2011; USCC Annual Report 2008). This type of attack is meant to paralyze an opponent through a swift and surprising strike, comparable to pressure points used in the martial arts, or the concept of an assassin's mace. Assassin's mace, or shashoujian, is used in Chinese military writings to describe a weapon or tactic "which can deliver decisive blows in carefully calculated surprise moves and change the balance of power" (USCC Annual Report 2002; Krepinevich 2010). Similar concepts can be seen throughout China's history, from Sun Zi's (tr. 1963) *The Art of War* to Mao Zedong's (tr. 2000) *On Guerrilla Warfare*. An assassin's mace enhances its strength by ignoring pre-established rules of conduct. It has many similarities to asymmetric warfare, such as being a novel way to deal with a more powerful adversary, but it differs in that it is a decisive weapon, aimed at incapacitating an enemy, "suddenly and totally" (Navrozov 2005). China possesses several asymmetric, highly devastating weapons, such as a limited but modernizing nuclear triad capacity, China's ASAT capability, and an electromagnetic pulse (EMP) capability. However, each of these has considerable drawbacks.[12] For example, human rights and environmental concerns have largely relegated nuclear weapons to the role of deterrent and introduced limited warfare. By using CNA, China could theoretically achieve the asymmetric destructive power of ASAT or EMP weapons while bypassing many of their drawbacks. Being swift in such a cyber attack is not only effective, it may be a requirement for success. If an opponent has time to prepare, they could take defensive measures, such as disconnecting critical systems from the Internet. This need to strike early in a conflict could sway military commanders to take pre-emptive action. Embedded backdoors (discussed further in the Computer Network Exploitation subsection below) would enable a swift retaliatory response against critical systems, and it fits the policy of active defense as a means of being prepared to counterattack.

All computers and networks that are connected to the Internet can become targets, and an attacker can gain up to total control of these systems. Therefore, theoretical examples of potential targets, types of attack, and tactics are vast. For example, targets which are of plausible value include corporations, energy distribution, financial institutions, government agencies, news organizations, transportation departments, universities, and many more—all of which are broad categories in themselves composed of numerous valuable targets which would require varying combinations, tactics, and outcomes. An attack can damage or destroy computers, such as remotely updating the Basic Input/Output System (BIOS) firmware with corrupt or defective data—the data which is necessary for booting a computer (Ingersoll 2013). Other methods of attack can temporarily deny a computer's use through harassment, such as DDoS, continually turning monitors off, or altering keyboard and mouse functions. Data on a computer can also be damaged, while leaving the hardware operable, such as deleting or encrypting valuable files (Taylor, Fritsch, and Liederbach 2015, 29). This wide range of potential targets and CNA-based effects yields a plethora of theoretical possibilities. States do not release detailed cyber contingency plans, because it would spoil their effectiveness and create conflict. Further, researchers attempting to catalogue all possibilities and devise the optimal plan of attack for various scenarios could risk being viewed as a threat to domestic or foreign state security. For these reasons—scarce use of CNA by states up to the present, the large range of credible possibilities, and limited open source research on the topic—a comprehensive examination of this aspect of Chinese CNA is beyond the scope of this book. This chapter will, however, attempt to reveal general trends, key examples, and some novel applications.

While surgical strikes may be preferable in CNA, China would likely cultivate a scattershot approach as well. Both of these possess benefits, and the two could be used in combination. The diversity and quantity of computer networks would appear to necessitate careful target selection. For example, a state attempting to impact the outcome of a war might receive little benefit from attacking dispersed individuals, banks, companies, or websites. These systems are diverse and redundant, which would seem to require significant resources to identify and exploit the vulnerabilities in a large enough quantity to have an effective impact. Embedded backdoors could be useful in this regard as a preparation for attack, although even this would benefit from targeting key organizations, or software and hardware that is in wide circulation, to increase effectiveness. However, given the apparent ease of CNE, and the large number of targets China has already allegedly exploited (see chapter 2), it is possible that they could plant latent kill switches across a wide range of targets in sufficient quantity to be effective. They could also employ online people's war to this end, for which China has the world's largest number of Internet users. Mobilizing China's population might not even be necessary to conduct such a large-scale operation. China could use botnets[13] to conduct such attacks. While attacks like DDoS and webpage defacements are often considered low-level attacks in terms of sophistication, they can still yield significant impact, including economic damage (Taylor, Fritsch, and Liederbach 2015, 29–31). For example, if China decided to take aggressive moves to settle maritime disputes surrounding the Paracel, Spratly, or Senkaku/Diaoyu Islands, such as establishing a stronger foothold in these regions, DDoS attacks could provide a timely disruption or distraction to deter international intervention, while also providing plausible deniability. The impact of this type of attack is comparable to a blizzard or flood; there is disruption and damage, but a city or state will likely recover in a relatively short time frame. This type of CNA could also serve as a coercive tool, or punishment apparatus, to encourage abiding by Beijing's wishes, particularly among its regional neighbors; and it could be used to draw two enemy states into conflict with each other. In a full-scale war, China could use DDoS and web page defacements to weaken an opponent's morale and the population's willingness to support government agendas, or it could disrupt communications and the functioning of society. In a technique similar to human shields, the attacks could be routed through the networks of foreign hospitals and schools to hinder retaliatory strikes which could stop the attack. With the ability to obtain remote control of computers, China could even use an opponent's computers to attack the opponent. Lastly, a host of opportunities exists for combining these attacks with conventional warfare to create a synergy. For example, emergency response communications and transportation systems could be disrupted, and disinformation, such as misdirection, could be planted, thereby hindering the disaster relief effort and herding people together to maximize the destruction caused by conventional weapons. A holistic approach to the development of CNA is to China's benefit, and this appears to be taking place through China's support of militias, quasi-support of hacker groups, and references to online people's war.

One specialized category of networks, which would likely fall under the scattershot CNA approach, is mobile phones. Ownership of smart phones, which commonly feature Internet connectivity, surpassed one billion users in 2012 (IDC 2014; Reisinger 2012). As with standard computers, their rapid growth has given rise to new vulnerabilities. China itself reported 702,800 mobile phone malware[14] infections in 2013 (AP-CERT Annual Report 2011). With mobile phone malware an attacker can disable a phone, delete data, dial incriminating or embarrassing phone numbers, replace the icon and wallpaper images, secretly enlist the phone into a botnet, send costly text messages or phone calls (billed to the victim), and send harassing messages (including to the victim and those listed in the victim's address book). Mobile spyware also exists, although eavesdropping and stealing data belong to exploitation rather than attack. Mobile phones can be infected through application (app) downloads, e-mail attachments sent to the phone, Multimedia Messaging Service (MMS), or standard computer-based methods when the phone is connected to the Internet or a personal computer. They can also be infected via Bluetooth wireless technology. For example, phones infected with the 2005 Commwarrior worm attempt to spread the worm by connecting with nearby Bluetooth-enabled phones. The target user must accept the incoming download to become infected; however, if they decline, the message will continue to instantly pop up until the user either accepts the download or moves out of the wireless range of the sender (Hypponen 2006). Again, China could use these types of CNA to demoralize an opponent and create civil unrest, or to reduce an opponent's ability to communicate and coordinate actions. Furthermore, it could be used to erode consumer confidence in non-Chinese products.

Despite the apparent development of rudimentary, large-scale CNA that would target more indiscriminately, authoritative PLA writings on CNA appear more focused on acupuncture-style warfare and targeting critical nodes. Interestingly, some of these critical nodes possess unique architecture and means of communicating with the Internet, such as wireless connections and quasi-closed networks. Among these are cloud computing, Point of Sale software and hardware, and a large number of satellite networks. For example, cloud computing possesses a distinct, yet online, architecture and large amounts of data concentrated at a single source. Cloud computing is designed to store some of an individual's information and applications on remote servers rather than on the individual's computer. This allows the individual to access these resources from any computer by connecting to the server; it also allows for shared resources which can reduce the operational costs of organizations. However, the majority of security concerns associated with this form of computer network primarily relate to computer network exploitation, and defense against exploitation, not attack. Examples include concerns of data theft, or concerns that companies basing their cloud computing operations in China could be coerced through China's legal apparatus to reveal network vulnerabilities and trade secrets. The attack aspects of cloud computing do not appear distinct from those found in common Internet architecture. The attraction of cloud computing in terms of cost savings could lead to increased concentrations of data being located in clouds, thereby making them critical nodes and appealing targets. This is ironic, given that cloud computing is meant to enhance an individual's computer security through backup data in the cloud in the event that their computer is attacked. By being part of a larger group, they become a more attractive target, because more damage could be inflicted. Similarly, for CNE, it is comparable to robbing a bank, rather than robbing individuals.

Another variant network which could be targeted by CNA is Point of Sale (POS) systems. These systems use wireless communication and Internet connections in portions of their network. The reasons for this are efficiency, productivity, and cost. Retail stores may need to communicate with the corporate network, vendors, and "external credit card processors" to validate transactions (Trend Micro 2014). POS software and hardware also suffers from a lack of vendor diversity and poor implementation of security features (Fritz 2008, 66–67; Raz 2013). Stores which use these wares are critical nodes, as they can provide an attacker with large amounts of data traffic from one source, particularly if access to the larger corporate network can be obtained. Between 2005 and 2007, a group of 11 people managed to steal 45 million users' bank and credit card details, resulting in a loss of more than $256 million. Their unprecedented feat was accomplished by parking outside of TJX retail stores and hacking into the store's wireless network. Once the account and pin numbers were obtained, the group was able to imprint this information onto blank bank and credit cards, including the magnetic stripe. These cards were then sold online or used to withdraw cash from ATMs. They would go to ATM machines near midnight, so that they could withdraw the maximum daily limit from each card, and then after midnight, withdraw a new day's maximum daily limit. The group's mastermind, Albert ("soupnazi") Gonzalez, was working for the U.S. secret service at the time of this crime; he had been recruited for his computer skills that were demonstrated in prior criminal activity for which he was apprehended (CBS 60 Minutes 2009; Verni 2010; Vijayan 2007).[15] Attacks targeting POS transactions, such as another breach in 2013 which stole bank and credit card account details from 40 million Target Corporation customers, are on the rise (McCoy 2013). While this would appear to appeal more to criminal organizations in search of profit, the societal disruption and cost to business competitors could appeal to a state. In particular, targeting the sale and distribution of a critical commodity such as petroleum could cause advantageous disruption during a time of war. Further, the tools and methods which allow access to POS software and hardware could be used to disable networks rather than steal from them.

A third type of network with unique architecture, which could be targeted in CNA, is satellite networks. Several satellite providers utilize wireless communication in portions of their Internet-connected networks, including radio and microwave signals used for uplinks and downlinks. Satellites are vital to sustaining the current balance in the global economy, society, and advanced militaries. As such, states are increasingly recognizing satellites as critical infrastructure. They provide a significant role in climate and natural disaster monitoring, communication, early warning systems, global broadcasting, meteorology, navigation, precision strikes, reconnaissance, remote sensing, surveillance, and the advancement of science and technology. Satellite services also have a supporting role in "mobile and cellular communication, telemedicine, cargo tracking, point-of-sale transactions, and Internet access" (GAO Critical Infrastructure Protection Commercial Satellite Security Should Be More Fully Addressed 2002).[16] The global commercial space industry alone has "estimated annual revenues in excess of $200-billion" (Space Security Index 2012). As of December 2012, there were an estimated 1,046 operational satellites belonging to 47 states in addition to various international entities and collaborations (UCS Satellite Database 2012). Beyond states, 1,100 firms in 53 countries use outer space as part of their operations (Robertson 2011). One type of satellite infrastructure, the very small aperture terminal (VSAT), is commonly used for bank transactions between headquarters and branches, Internet access in remote locations (including Intranet, local area networks, video conferencing, virtual private networks, and VOIP), mobile or fixed maritime communications (for example, a ship or oil rig), POS transactions, and Industrial Control Systems. A significant disruption to satellite services would have damaging effects on society. In addition to Internet websites and mobile apps that reveal particular frequencies used by individual satellites, scanning software exists that can automate the process for potential hackers (Laurie 2009). The four primary types of satellite hacking are Jam, Eavesdrop, Hijack, and Control. Attempts to increase security through encryption, hardening, and redundancy all carry an increase in cost and a reduction in efficiency which commercial satellite providers may not see as worth the risk reduction. A 2004 study published by the U.S. President's National Security Telecommunications Advisory Committee "emphasized that the key threats to the commercial satellite fleet are those faced by ground facilities from computer hacking" (Space Security Index 2012). A reliance on satellites can be viewed as a soft spot which could be exploited. For example, authoritative "Chinese military writings advocate attacks on space-to-ground communications links and ground-based satellite control facilities in the event of a conflict" (USCC Annual Report 2011).[17]

Critical infrastructure, including the quasi-closed networks of ICS, are also vital nodes which would fall under China's acupuncture war. These include chemical plants, emergency services, energy grids, communications, production lines, transportation systems, waste treatment, and water utilities. Such attacks could cripple the flow of goods, cause panic, instigate civil unrest, create food shortages, and shut down business. Non-state actors have proven that the capability of such an attack is plausible. For example, cyber attacks which have been conducted against these types of key infrastructure include: the disruption of emergency response by embedding malicious code into e-mail; disrupting air traffic control communication, including the ability "to activate runway lights on approach"; using a worm to corrupt the computer control systems of a nuclear power plant in Ohio; and using a Trojan horse[18] to gain control of "the world's largest natural gas pipeline network" (DCSINT 2005; Lourdeau 2004; Wilson 2003; Wilson 2008; Maynor and Graham 2006). In 2000, Vitek Boden used radio transmissions to remotely trigger the release of nearly a million liters of sewage into the public waterways of Queensland, Australia (Abrams and Weiss 2008; Barker 2002). Power outages in Brazil in 2005 and 2007, which affected millions of customers, were attributed by some sources to cyber attacks (CBS 60 Minutes 2009; Lewis 2010). There have also been some alleged state-sponsored attacks targeting critical infrastructure. Despite widespread reporting by analysts and the media that the Stuxnet worm was "the first cyber weapon," the United States allegedly inserted a Trojan horse into computer systems being sold to the Soviet Union, which in 1982 caused an explosion in the Trans-Siberian Pipeline.[19] Further, the United States was allegedly engaged in multiple programs during the 1980s to embed backdoors into computer products for enhancing CNA and CNE (Markoff 2009; Safire 2004; Weiss 2007). In 2007, Israel was also accused of using pre-installed backdoors, in this case to remotely disable Syrian radar systems (Adee 2008). As noted elsewhere in this book, North Korea and Russia have both been accused of conducting low-level CNA, such as DDoS and web page defacements on multiple occasions, and in 2012, Iran was accused of attacking the Saudi Aramco petroleum and natural gas company. The attack on Saudi Aramco used a virus to erase data on internal networks, such as "documents, spreadsheets, e-mails, [and] files" and leave behind "an image of a burning American flag" (Perlroth 2012).

The United States and Western countries are particularly vulnerable as much of the communication, manufacturing, water, transportation, and energy infrastructure is owned by the private sector, as opposed to China and Russia where infrastructure is predominantly in the hands of the government (Greenemeier 2007). U.S. government and defense installations are heavily funded for security, whereas the private sector is not. Initially, the U.S. power grid control systems were on closed networks (not connected to the Internet); however, as organizations grew, and so did the Internet, it became more cost effective to tie them together. In particular, with deregulation it became more important for offsite maintenance and information sharing. The Internet became essential for operations, meaning they would need two separate systems for operation, one connected and one not. Through the decision-making process companies decided it was cheaper to have only the one that was connected, but focus on keeping it secure. Over time security became lax, and no network that is connected to the Internet can be entirely secure. Many of these systems do not support authentication, encryption, or basic validation protocols; of those that do support them, most run with security features disabled (Maynor and Graham 2006). Further, power companies may buy and trade power among themselves, so loopholes designed to check available capacity have provided another entry point (Winkler 2007). In addition to Internet connections, ICS may be compromised through outdated modems used for maintenance purposes, wireless access points, roaming notebooks, and embedded pre-installed exploits in software and hardware. For example, spear phishing[20] targeted at individuals who have access to a closed network could lead to the installation and spread of a virus on an open network. This virus could then be carelessly transported on removable data storage between the open and closed network. The widespread use of data storage devices, such as USB flash (thumb) drives, cameras, mobile phones, mp3 players, and portable gaming devices, makes their presence in sensitive locations appear innocuous. The vulnerability of the private sector's computer networks, due to a lack of understanding or a lack of incentive, provides China (or other cyber-capable groups) with the opportunity to cripple U.S. infrastructure. In some cases, reports of the vulnerability of ICS may be alarmist,[21] and media portrayal sensationalized,[22] to support particular agendas; however, given the known instances of CNA targeting critical infrastructure, coupled with widespread allegations of state-sponsored probing of these networks, it appears to be a credible threat. In the case of the energy grid, striking a small number of key nodes could create a cascading failure.[23] Further, a state would have considerably more resources to devote to CNA development than the non-state actors who have already conducted such attacks. Lastly, the target of such an attack could be any state, not only cyber-advanced states, which raises its feasibility as an attack option.

The final specialized network and critical node to be examined in this subsection, which will help to round out the scope and complexity of potential CNA, is the United States' Nonsecure Internet Protocol Router Network (NIPRNet). Cyber warfare analysts suggest that this may be a prime target in the event of a conflict between China and the United States (Mulvenon 2013). NIPRNet contains unclassified, yet sensitive, information pertaining to the U.S. government and military which are critical in times of peace and war. This information includes,

> all DoD bill payments; the daily calendars for admirals and generals; troop and cargo movements; aircraft locations and movements; aerial refuelling missions; medical records for military personnel and their dependents; soldier and officer evaluation reports; unit deployment information; and all e-mails among Department of Defense and military personal digital assistant communications devices. (USCC Annual Report 2008)

This quasi-closed network maintains connections to the Internet for ease of use and enhanced productivity, including providing access to those in remote locations. China has been accused of accessing and altering data on NIPRNet in 2006 and 2007, including implanting malware in the network which could have been used in an attack (Krekel 2009; Tkacik 2008). One of the functions of NIPRNet is the coordination of force deployment, including non-combat aspects, such as the transport of supplies and supporting personnel which are essential to operations. In times of an impending war with a distant opponent, such as the Iraq War in 2003, the United States begins operations with a large buildup or deployment phase prior to engagement, actions which are widely reported on in the international news. If China decided to reclaim Taiwan in a surprise move, it could target NIPRNet to delay U.S. involvement, and potentially obtain Taiwanese surrender before the United States could fully respond.[24] The network could be sabotaged, or made temporarily unavailable through a DDoS attack, or important data could be deleted. More subtly, China could alter data on the network to disrupt logistics. For example, equipment, supplies, and personnel could be sent to the wrong locations, and coordinates or supply lists could be tampered with (China's Proliferation Practices 2008; Lewis 1994). Using CNA to target NIPRNet could avoid direct confrontation and provide plausible deniability.

COMPUTER NETWORK EXPLOITATION

While chapter 2, *The Development of Computer Network Exploitation, 1998–2014,* provides a comprehensive examination of alleged Chinese CNE incidents, it is relevant to discuss it here, as it is part of CNO. The specific incidents are sufficiently extensive and instructive to merit their own chapter.

Computer network exploitation (CNE) refers to the clandestine theft of information through Internet-connected hacking; and it is synonymous with the terms cyber espionage and cyber reconnaissance. The online theft of intellectual property allows China to leapfrog in research and development across all sectors. Not only does this theft allow China to advance its own capabilities, it also erodes the power of victim states. Companies, militaries, and governments might invest heavily in time and money to produce products, but lose the reward of a competitive edge when those products are stolen. Furthermore, money and efficiency are lost through attempting to improve computer security against CNE. In another form of information theft, CNE can eavesdrop on activist activity, business deals, enemy weaknesses, impending news, government intentions and sentiment, and military developments and capability. Unlike traditional espionage, CNE is capable of global reach at high speed.

Intruders can also operate anonymously, and at a relatively low cost, from the safety of their own country.

The following template is allegedly a common method used by China to conduct CNE (HBGary 2011; Lawrence and Riley 2013; Mandiant 2013; Stokes and Hsiao 2012; USCC Annual Report 2009). First, the Internet is used legally to identify and research individuals and organizations which may possess information that is useful to China. A spear phishing campaign is then conducted against these individuals using information gathered from company websites and social media. The aggressor attempts to tailor the e-mail content to the target, so it appears as if it is arriving from a legitimate source.[25] The bait message might include upcoming meetings, products, and events, or associate and friend names, as subject matter. The e-mail address used to send the message can also be crafted to appear innocuous; for example, selecting an e-mail address from a free e-mail service provider that closely resembles the local and domain components of the target organization's employee-only addresses. Malicious attachments and links can also be disguised, such as changing the file icons, hiding file extensions, or using misleading display text for a link rather than displaying the full URL. Clicking on the link or attachment within the deceptive message can run a harmful program (if it is an executable file) or take advantage of a wide range of exploits, including zero-day exploits,[26] to obtain varying levels of unauthorized access to the computer. This access can be used to enhance additional spear phishing against further targets within the network or organization. For instance, by combing through the e-mail account of a compromised target, the intruder can learn the common structure of internal messages, the victim's contact list, and current topics; and the intruder can begin sending malicious mail from the victim's address, thus increasing its perceived legitimacy. Moreover, they may have access to all of the files on that computer, or access to the local area network (LAN), which can yield additional information, including passwords for other accounts and services the victim uses. Among the tools which can be installed in the initial infection is a remote access tool (RAT),[27] which can allow the intruder total control over the infected computer. With complete, or even partial control, the intruder can install additional payloads[28] beyond the initial infection, such as a keylogger[29] or alternative backdoors for remote access. The initial breach may be eventually detected and repaired, so it is important to establish secondary entry points before access is cut off.

Adding to the common template of alleged Chinese CNE, the commands being sent to control computers that have been infected by spear phishing are routed through intermediary computers, often in foreign countries, to combat detection and identification. For instance, China may have control of compromised computers in Japan and Taiwan[30] and use these as the command and control (C2) computers for conducting CNE against the United States. Relaying both the instructions and stolen data through an intermediary aids anonymity, because the victim would see the intrusion as coming from these intermediaries. Typically, a computer can only identify the IP address[31] of computers attempting to connect with it, not the IP address of third parties which may be issuing that order. In other words, each computer can only see the next link in the chain. The aggressor can see in both directions, because they have taken control of the intermediary, which the victim has not. This can be illustrated through the aforementioned countries: even if the United States were to collaborate with Japan and Taiwan to determine that the intermediary was controlled by computers in China, China could claim that its computers were also under remote control and stymie further investigation efforts. Spear phishing is not the only entry point of alleged Chinese CNE. Other entry points include password cracking, scanning for vulnerabilities,[32] and watering hole attacks[33]; however, spear phishing is one of the most frequent methods (Stokes and Hsiao 2012). Intruders use this initial access to move laterally in a network and escalate privileges. Once a network has been breached, it is scanned, literally and figuratively, for data of value. This data is often moved to staging areas within the compromised network, where it is compressed into zip or rar file formats and encrypted for fast and secure export. By transporting the valuable data to a staging area, the intruder can reduce their "operational footprint," spend less time in sensitive areas which may be closely monitored, and blend in with normal LAN and Internet traffic activity (USCC Annual Report 2009). Alternatively, exfiltrating data outside of business hours in the host country can aid detection avoidance as there would be fewer personnel present to notice large flows of traffic. This is assuming the modems remain powered on during non-business hours. Compromised computers, on the other hand, can be remotely programmed to turn on at a specified time by the intruder.

Some CNE activities which test the boundaries of its definition include mapping systems for vulnerabilities and setting up for computer network attack. In many cases an intruder can switch from exploitation to attack at will, since both largely rely on the same tools and skillsets, and both offer limited or total remote control of a victim's computer; it is up to the intruder to decide the purpose for which control was obtained. Therefore, networks which have been penetrated in exploitation, from energy distribution to financial services and hospitals to transportation systems, can be viewed in a way comparable to "radar lock-on"; the intruder had a "clear shot" but chose not to fire. However, once an attack is made, it is unlikely that an intruder can retreat to exploitation, because they will have revealed their presence and possibly damaged the system in a way which no longer permits the collection of information. An attack carries greater consequences, as evidenced by China's tentative agreement that the law of armed conflict applies to the cyber realm, and evidenced by the sparse amount of state-sponsored CNA allegations compared to the vast amount of state-sponsored CNE allegations. For China, which in many ways remains a developing country, Western countries are lucrative CNE targets. If China were to use CNA, it could risk upsetting the benefits it can gain from CNE, and it could face retaliatory attacks. While China may have less to lose in tit-for-tat intellectual property theft, it does have a great deal to lose in terms of the destruction of computer networks. This does not mean that conducting CNE is without consequences for China either. In addition to risking international condemnation and escalation of tensions within and outside of the cyber realm, China also risks setting a precedent for CNE. While it remains a developing country, China's rapid advancements have seen it become a leader in many categories; including the world's second largest economy, the third nation to conduct independent human spaceflight, and the fifth nation to obtain nuclear weapon capability. China's views on the morality of intellectual property theft may change, if it becomes a primary target of other developing countries. As an example, Brazil could greatly benefit from the trade secrets of China's space program or telecommunication companies.

A final component of CNE is the use of embedded backdoors. These are pre-installed exploits in hardware, particularly microchips, although the term also applies to software, such as "implanting malicious code" during the manufacturing process, which "could be remotely activated on command" (USCC Annual Report 2008). In the case of China, these are sometimes referred to as Manchurian chips or Trojan dragons. Embedded backdoors can also be used for CNA; however, their presence alone does not necessarily reveal intent. Concerns over China's potential use of embedded backdoors appears to have gained prominence in mass media around 2008, and particular scrutiny was placed on the Chinese telecommunications company Huawei in 2012 (CBS 60 Minutes 2009; Tkacik 2008; USCC Annual Report 2012). Yet, accusations against Huawei date back to 2003, and the concept of embedding backdoors is not new to Chinese CNO strategists (McDonald and McGuirk 2012; Rogers 2012). In a 1988 lecture at the Chinese National Defense University, Dr. Shen Weiguang, a former PLA officer, State Council member, and prolific author on information warfare (IW) theory stated:

> Virus-infected microchips can be put in weapon systems; an arms manufacturer can be asked to write a virus into software; or a biological weapon can be embedded into the computer system of an enemy nation and then activated as needed. Thus, war preparation takes on another form. Preparation for a military invasion can include hiding self-destructing microchips in systems destined for export. (Thomas 2004, 45)

To this end, China may seek to establish market dominance in the production of information and communication technology (ICT) software and hardware. If China could unseat Microsoft as the industry standard in software, it could install backdoors, latent viruses, or remotely triggered ex-filtration devices. In this case CNE could create a positive feedback loop. CNE could help enable China to obtain market dominance through industrial espionage, and once market dominance is obtained, it could enable greater CNE capabilities through embedded backdoors. Moreover, market dominance would boost China's economy, and it could aid Chinese CND by limiting other countries' ability to use pre-installed exploits. Lastly, China could use its economic might to obtain greater ownership of submarine cable infrastructure. This would allow further access to cyber reconnaissance or the option of shutting down portions of Internet connectivity during times of conflict.

COMPUTER NETWORK DEFENSE

All computers which are connected to the Internet are susceptible to attack and exploitation. There is no guaranteed defense—other than unplugging from the Internet, which may defeat the computer's intended purpose—there are only ways to mitigate risk. The Internet plays a critical role in China's economy, society, and modernization goals; and therefore it is a strategic asset which must be protected. In 2012, 676 million devices were used to connect to the Internet by China's 538 million Internet users (USCC Annual Report 2012). Given that China has the world's largest Internet population, in terms of volume China has the most targets to defend. PRC President Xi Jinping referred to information technology as a "double-edged sword" (Pace 2013). On the one hand it offers a new means of prosperity, and on the other it creates new security vulnerabilities. The more society becomes reliant on technology, the greater the cyber security risk; yet to maximize the benefits technology offers, including profit, requires a full commitment. For example, it would be inefficient to maintain all of the old paper-based record keeping and transaction infrastructure as a fail-safe for new electronic systems. A reluctance to embrace technology and globalization can be detrimental, as demonstrated in Chinese history. The naval expeditions of Zheng He during China's early Ming dynasty had secured China's position as a great power; however, a change in leadership and a turn toward xenophobic policies led to a decline in Chinese power. In particular, lagging naval technology played a role in China's Century of Humiliation, and China is now attempting to regain a top-tier naval power status. China must embrace the Internet to achieve its goals, despite the new risks it places on state security, social stability, and sovereignty,

In 2008, industries associated with the Internet accounted for 1.6 percent of China's GDP, and 10 percent of its global trade (The Internet in China 2010). Unlike industrial manufacturing, this industry meets China's goals of reducing carbon emissions and producing goods higher on the technological ladder. Further, Internet-related industries provide a means for boosting soft power[34] and the spread of Chinese culture through online animation, apps, gaming, memes, music, and video. China is seeking to develop world-recognized brand names in hardware, software, and services, comparable to what the United States has done with companies like Apple, Google, Facebook, Microsoft, and Yahoo. Beyond attempts to reach a global audience, the Internet has become integral in Chinese society. According to the Information Office of the State Council of the People's Republic of China, in 2009, 14 million Chinese people used the Internet to organize trips, 15 million searched for employment opportunities, 35 million participated in securities trading, 100 million purchased goods online, and 230 million sought information through search engines. Further, in 2009 there were 45,000 government websites alone, and 80 percent of online users relied on the Internet as their primary source of news (The Internet in China 2010). China's Twelfth Five Year Plan (2011–2015) calls for the enhancement of broadband access penetration, cloud computing, electronic payments, and mobile communication; the latter two being facilitated in part by Internet connectivity and shown in the CNA subsection of this chapter to be vulnerable nodes. Further, the Twelfth Five Year Plan calls for the construction of digital and wireless cities, an employment information database, and intelligent power grids; along with the convergence of the Internet, radio, telephone, and television. China is also attempting to increase operational efficiency by promoting the use and interconnection of computer networks in all aspects of business and government, the distribution of goods and raw materials, the health care system, and real estate supervision and standardization (China's Twelfth Five Year Plan 2011). All of these point toward increasing reliance on the Internet in China, which elevates the importance of computer network defense.

Along with the dramatic growth of the Internet in China, there has been a dramatic growth in Internet-related crime targeting China; however, government and media reporting on this issue remains vague in open source documents. For example, following international criticism of alleged Chinese CNE in 2007, the Vice Minister of the Ministry of Industry and Information Technology (MIIT) of the PRC stated that China has also been a "victim of massive and shocking losses of state and military secrets via the Internet" (Leyden 2007). Moreover, in response to a high-profile and detailed report released by the cybersecurity firm Mandiant in 2013, the director of the National Computer Network Emergency Response Technical Team/Coordination Center of China (CNCERT or CNCERT/CC), Huang Chengqing, denounced Mandiant's claims and stated that China has "mountains of data" if it wished to accuse the United States (China has "mountains of data" about U.S. cyber attacks: official 2013). CNCERT does produce weekly and annual cyber security statistical data which provides a fuller picture of China's CND situation, albeit with some uncertainties. The 2013 CNCERT annual report claims that CNCERT is a non-governmental organization, yet previous CNCERT annual reports claim it is "under the leadership of MIIT" and provides support to government departments (APCERT Annual Report 2011; CNCERT/CC Annual Report 2006). Further, CNCERT's reported data on the number of victims of computer network attack and exploitation includes foreign organizations operating in mainland China. For example, the "Top 10 Phishing Victim's" for 2009 are all foreign organizations, many of which have been reported elsewhere as victims of Chinese state-sponsored CNE (see chapter 2) (APCERT Annual Report 2009). Director Huang asserted that foreign allegations of Chinese state-sponsored CNE are unconstructive, and some cases could have been solved if the victims had presented their intrusion-related information to CNCERT (China has "mountains of data" 2013). Given the unclear ties of CNCERT and the Chinese government, this could be the equivalent of sharing what is known about the intrusion with the intruder, and then being added to statistical data as a Chinese victim. Adding to the vague nature of CNCERT reports, there is no detail given on collection methods, level of attribution, or the difference between category titles, such as backdoor, malware, phishing, and vulnerability, which would seem to have significant overlap. Despite these limitations, the CNCERT reports are official open source documents supported by the PRC, and they form a part of the broader, multination, Asia Pacific Computer Emergency Response Team, so there does appear to be some accountability and quasi-international endorsement.

CNCERT received 4,390 incident reports in 2007, 10,433 in 2010, and 31,700 in 2013 (CNCERT/CC Annual Report 2007; CNCERT/CC Annual Report 2010; CNCERT/CC Annual Report 2013). The primary means of exploitation or attack during these years were spear phishing, web page defacement, web page–embedded malicious code (possibly watering hole attack), and the nondescript terms vulnerability and malicious code.[35] Beyond the incidents that were reported to CNERT in 2013, CNCERT itself tracked 29,100 overseas C2 servers, 30,200 "phishing sites" targeting mainland China, and 47,300 Chinese IP addresses that were under illicit remote control.[36] In all three of these types of malicious activity—C2 servers, phishing, and remote access—the United States is alleged to be the largest non-Chinese perpetrator, as is consistent with previous CNCERT annual reports. Other countries cited in multiple CNCERT annual reports as the source of malicious activity are India, Japan, South Korea, and Taiwan. Additionally, in 2013, 24,000 web pages were defaced, including 2,430 government websites. This is a decrease from the 34,845 in 2010 and 61,228 in 2007; however, web page defacements are becoming increasingly frequent when viewed over a longer period of time. Internet-enabled mobile phones suffered 6,249 malware infections in 2011; 163,000 in 2012; and 702,800 in 2013. CNCERT began reporting on mobile infections in 2010, citing more than three million mobile infections in that year. The sharp decline in 2011 could be due to increased protection or a change in mobile phone design, or it could represent a change in the process of monitoring and reporting infections. Despite the sharp decline in 2011, mobile infections have been steadily increasing since then, and the large number of infections in all four years represents a significant vulnerability which requires defense. Sporadic reporting of CNA and CNE statistics from Chinese news outlets and government agencies vary slightly from the CNCERT reports, yet concur with the overall trends—annual attacks, infections, and intrusions are in the tens of thousands, they are increasing, and a large percentage originate from foreign countries, particularly the United States (China's cyber security under severe threat: report 2013; China says hit by 500,000 cyberattacks in 2010 2011; China says U.S. routinely hacks Defense Ministry websites 2013; Li 2011; USCC Annual Report 2012). According to the report *China's 2010 Corporate Security Threats*, released by the Internet security company Beijing Rising Information Technology, classified Chinese networks were targeted by CNA and CNE more than 10 million times in 2010 (Henderson 2011). Intruders attempted to use removable media, including mobile phones, as entry points into the classified, and possibly air-gapped, networks. The London-based market intelligence company Economist Intelligence Unit ranks China's computer network defense ability as "13th out of the G-20 countries" (USCC Annual Report 2012).

As illustrated by the select computer network attack examples earlier in this chapter, non-state actors can cause an immense amount of damage to a state, stealing information, deleting and changing files, transferring capital, and destroying programs or entire networks. Gary McKinnon was a foreign solo hacker, Vitek Boden was a domestic insider, and the TJX incident was conducted by a combination hacker group and criminal organization. As another example, Aum Shinrikyo[37] represents a cult and terrorist organization. The rapid advancements in technology and globalization have given individuals the capability to cause massive damage with nothing more than an off-the-shelf computer and an Internet connection. Non-state actors using computer network operations can threaten everything from business and government to military and critical infrastructure. According to Joe Weiss, an industrial control systems specialist and author of *Protecting Industrial Control Systems from Electronic Threats*, China has suffered at least three incidents targeting critical infrastructure, none of which were attributed to state-sponsorship and one of which involved insiders. The first concerned Ertan Hydro Station's control system which "received unexpected signals." This was followed by a near system collapse when "it reduced generation 900 MW within 7 seconds." Only through disconnecting the hydro station's control system from the Internet could the situation be salvaged. The second instance of critical infrastructure being targeted was the installation of "logic bombs"[38] by disaffected employees "on more than 140 disturbance recorders causing malfunctions." The perpetrators were arrested. The third episode involved "viruses [that] were found in three high-voltage direct current (HVDC) converter stations (Longquan, Zhengping, and Ercheng) that transfer a total 6000 MW from the Three Gorges Dam" (Weiss 2010, 144). In the civilian realm, China is renowned for its use of pirated software, which often contains preinstalled malware and lacks "important security updates" (Wilson 2007). As noted in the CNCERT annual reports, and further detailed in chapter 2, China has suffered numerous web page defacements and DDoS attacks from hacker groups. Among these were attacks in 2012 which were attributed to the group Anonymous (Anonymous says it hacked Chinese government sites 2012; Anonymous says it will hack more Chinese sites 2012). Anonymous is a large and diffuse group, which has been blamed for attacking a wide range of targets for diverse, and at times conflicting, motives. Given Anonymous' combination of ambiguity and notoriety, they (along with others) could serve as a cover for state-sponsored CNA against China.

There are an estimated 120 states developing cyber warfare capabilities with at least 10 possessing an advanced offensive capability (Brodkin 2007; Hopkins 2011; McAfee 2009; Paget 2013). Indeed, the international push for developing CNO capabilities has come to resemble what might be termed a "cyber arms race." It has been set into play through a combination of volatile global conditions, including the perception of a China Threat,[39] concerns over U.S. hegemony,[40] and an aggressively resurgent Russia.[41] China's 2013 National Defense White Paper directly acknowledges this competition, stating that great powers are "vigorously developing" their CNO capability in order to preserve "strategic superiorities" (Information Office of the State Council of the People's Republic of China 2013). As evidence of this, the United States and Israel have been accused internationally of developing the Stuxnet worm which targeted uranium enrichment facilities in Iran (Baldor 2013). Further, the United States' commitment to CNO was illustrated in 2013 with the proposed Distinguished Warfare Medal, which would have honored extraordinary achievement in cyber warfare or combat drone operations (Dwyer 2013; Shanker 2013).[42] Despite budget cuts in many U.S. military programs in 2014, the cyber operations budget increased by 21 percent to $4.7 billion (Hoffman 2014). Foreign states wishing to use CNO against the United States (or others) may recognize the international focus being placed on alleged Chinese CNE and use that to their advantage. They could use RATs to create botnets and C2 nodes in China, then use these computers to conduct reconnaissance and attack in other countries, leaving China to take the blame. This was demonstrated in the cyber attacks during the Russo-Georgian War of 2008. When Georgian officials attempted to block all Internet traffic coming from Russia, the attackers rerouted their attack through China (Clarke and Knake 2010, 19). Further, denouncements of alleged Chinese CNE may indicate that countries are developing retaliatory responses, and allegations of Chinese incursions could be used to bolster support for increasing CNO foreign budgets. These allegations in themselves are something which China must defend against, as it may affect China's economy by making investors cautious and export controls/legal bureaucracy more stringent. Similar to CNA and CNE being routed through China in an attempt to incriminate them, CNA and CNE against China can be routed through multiple countries to obscure the aggressor's identity—a tactic which China has been frequently accused of. On January 12, 2010, China's top search engine, Baidu, was attacked. Internet users attempting to access Baidu were redirected[43] to a web page which displayed the Iranian flag and the message "This site has been hacked by the Iranian Cyber Army" (Branigan 2010a). Several factors led to speculation that the United States was the true culprit: (1) it was unclear why Iranian hackers would target China, especially given their reciprocal support in energy deals and UN voting power; (2) the attack occurred on the same day in which Google publicly accused China of hacking into its computer networks[44]; and (3) the United States also had motive to frame Iran, in particular retribution for Iran's Internet censorship which blocked news and social media websites, including Facebook, Twitter, and YouTube, during the 2009–2010 Iranian election protests.

Classified documents leaked by former National Security Agency (NSA) employee Edward Snowden revealed U.S. CNE targeting China. Organizations targeted over a four-year period included Pacnet, a global telecommunications provider and owner of the region's largest fiber-optic submarine cable network, Tsinghua University in Beijing, The Chinese University of Hong Kong, and the text messages of multiple mobile phone service providers. Both universities targeted serve a dual role as Internet traffic hubs, or Chinese "network backbones," making all four targets exchange points "through which large quantities of data pass" (Lam 2013a; 2013b). Since 2007, the NSA has also allegedly targeted Chinese banks, government officials, the Ministry of Commerce, and the telecommunication giant Huawei, in an operation code-named Shotgiant (Sanger and Perlroth 2014; Targeting Huawei: NSA Spied on Chinese Government and Networking Firm 2014). Beyond the damage caused by the transfer of intelligence, these incidents create a financial cost for China as they attempt to investigate the breach, halt its continuation, and prevent future incidents. It also results in a loss of efficiency as former operating procedures can no longer be viewed as secure.

Despite mistrust and competition among states, CND offers some avenues for international cooperation, such as combating spam and select botnets. Non-state actors, particularly criminal and terrorist organizations, using CNO to enhance their activities are also a mutual concern.

China's Golden Shield content control mechanism (see chapter 3 for further detail) is derided by Western states, yet it possesses some security features which are desirable to Western states such as controlling hacktivist activity. The 1999 Seattle WTO protests, the 2010 G-20 Toronto summit protests, and the United States' PRISM surveillance program point toward some overlapping interests. Moreover, through CNCERT, China is an operational member of the 18-nation Asia Pacific Computer Emergency Response Team (CNCERT/CC Annual Report 2013). China has information security agreements with ASEAN and SCO, and it has sent representatives "to more than 40 countries" to research "Internet development and administration" (The Internet in China 2010). Further, China has participated in the World Summit on the Information Society (WSIS), the Interpol Asia-South Pacific Working Party on Information Technology Crime, and the Sino-British Internet Round Table. In conjunction with the United States, China participated in the China-U.S. Joint Liaison Group, the U.S.-China Internet Industry Forum, and joint cyber war games organized through the think tanks the Center for Strategic and International Studies and the China Institute of Contemporary International Relations (Hopkins 2012). Lastly, China has continually pushed for UN oversight of Internet administration; although this may be because it favors China's long-term interests rather than a collective good.

China can use domestic and international law to bolster its computer network defense efforts. For example, China's authoritarian control and centralized structure, such as State Owned Enterprises, could help in mandating computer-related security standards. Western states struggle with convincing the commercial and private sectors to boost CND as they often do not see the risk posed by CNO as being worth the additional cost. Further, government monitoring might be viewed as a violation of citizens' privacy and face strong opposition in the West. China has pursued several laws which require foreign companies seeking to operate, sell products, or register products in China to submit their proprietary information for inspection and approval (Heickerö 2012, 79; Ragland, McReynolds, Southerland, and Mulvenon 2013; USCC Annual Report 2010; USCC Annual Report 2011). On the surface this allows China to ensure the security of information technology, such as freedom from embedded backdoors, however it also has deeper implications. By enabling access to trade secrets, China can examine products for exploit vulnerabilities and reverse engineer foreign products. Reverse engineered products could conceivably be sold at a competitively reduced cost given the amount of money saved on research and development and the financial backing of the state. Additionally, companies that refuse to abide by China's regulations are not allowed to do business in China. The appeal of access to the Chinese market and the reduced production costs of operating within China provide incentives for foreign companies to abide by undesirable rules; even though these short-term profits may erode their long-term competitive lead.

For China, which is seeking to develop technology indigenously and boost the sales of domestic-made products within the country, this becomes a win-win situation—either foreign companies are blocked from the Chinese market because they do not meet China's conditions, or foreign companies allow access to proprietary information and are eventually driven out through a loss of competitive edge. As an example, video game consoles, defined as "computer entertainment system[s]," were banned in China from 2000 to 2013. The prohibition on video game consoles, such as Nintendo, PlayStation, and Xbox, was attributed to concerns over its effects on children, despite a thriving personal computer gaming market. In the buildup to a possible lift of the ban, producers of video game consoles who wished to conduct business in China were required to submit their products and supporting specifications in order to pass safety standards and obtain approval. As one example of possible reverse engineering, the Chinese computer company Lenovo released "a motion sensing device" in 2012, called an "exercise and entertainment machine," which appeared to be a copy of Microsoft's Xbox peripheral *Kinect* (Kelly and Lee 2012). This was not the first time China was alleged to have used legal and financial prowess to disadvantage Microsoft. Beginning in 2003 and continuing in subsequent years, China's National Development and Reform Commission, formerly known as the State Development Planning Commission, persuaded Microsoft to allow access to its operating systems' source code, as well as, provide education and training on manipulating the functionality of Microsoft software (Clendenin 2010; Li 2003; The State Development & Planning Commission 2002). Multiple sources, including a leaked 2009 diplomatic cable, indicate that this has provided China with an advanced CNE capability, particularly given that over 90 percent of personal computers globally use Windows operating systems (Desktop Operating System Market Share 2014; US embassy cables 2010; Tkacik 2008).

One of the difficulties with CNA and CNE is determining an appropriate response. If a state's communications center is attacked by a missile, it is considered an act of war. But what is the response to a cyber attack on that same installation, with the same debilitating effect? In 2013, China made news headlines by agreeing for the first time that the law of armed conflict applies to cyberspace; however, this was not as significant as it may appear. China was one of 15 states to agree on the contents of a UN report on *Developments in the Field of Information and Telecommunications in the Context of International Security*. Among the 34 non-binding articles in the report was the statement that the Charter of the United Nations is "applicable and essential" to the information communications technology environment (Group of Governmental Experts 2013). While not stated directly in the report, the law of armed conflict is tied to the

Charter of the United Nations. The law of armed conflict is itself open to interpretation, and Chinese officials do not appear to have made any public statement endorsing this position, while in the past they have stated their opposition (USCC Annual Report 2013). While the weight of this UN report is in question, it is important to note another article it contained which is particularly relevant to Chinese CNO:

> States must meet their international obligations regarding internationally wrongful acts attributable to them. States must not use proxies to commit internationally wrongful acts. States should seek to ensure that their territories are not used by non-State actors for unlawful use of ICTs. (Group of Governmental Experts 2013)

In addition to the difficulties of identifying the source of a cyber attack and determining an appropriate response, the absence of an international legal framework to deal with such an attack makes it exceedingly problematic.

CONCLUSION

The Internet-connected hacking of computer network operations (CNO) furthers China's trends of asymmetric warfare and technology transfer as a means of leapfrogging in modernization. China acknowledged the creation of an online blue army in 2011 composed of 30 operators, although Chinese source material not intended for a Western audience places the creation of China's first CNO unit between the years 2000 and 2003. China is believed to possess at least 20 units, as of 2016, with personnel numbering in the hundreds or thousands per unit. The General Staff Department is allegedly China's primary organization for conducting CNO, with its Third Department being responsible for exploitation and defense, and the Fourth Department being responsible for attack. While these are the primary organs for China's CNO, multiple entities maintain an auxiliary or independent role. These include technical reconnaissance bureaus, the four armed service branches, multiple government agencies, universities, militias, and hacker groups. As of 2008, China appeared to have 33 CNO militia and approximately 250 hacker groups, both having their origins in the late 1990s.

Computer network attack (CNA) is capable of destroying or damaging hardware; disrupting or denying service; and altering or damaging data through a variety of entry points, methods, and tools. All Internet-connected computers are capable of being penetrated, and an intruder is capable of obtaining up to full control of a computer. These factors combined mean there is an enormous range of possible targets, tactics, and outcomes. Beyond standard computer network architecture, some systems which are susceptible to CNA possess distinct means of transmission, such as cloud computing, industrial control systems, mobile phones, Point of Sale software, quasi-closed networks, and satellites. Critical infrastructure is a key target, as it appears Chinese CNA theory is geared toward striking critical nodes (nodes whose destruction or disruption would cause a great impact) in acupuncture-style warfare. To date, there are few state-sponsored CNA incidents in comparison to exploitation incidents. This could be the result of states wanting to hide their capabilities until they are necessary to achieve an imperative, or it could be that thus far the cost-benefit analysis has acted as a sufficient deterrent. Despite the relatively low number of incidents, there have been some, and non-state actors have further shown the scope and complexity of what is possible through CNA. Non-state actors have disrupted air traffic control, emergency response, energy grids, a gas pipeline, a nuclear power plant, and sewage systems. States have been accused of using CNA to destroy a gas pipeline, damage nuclear facilities, disrupt an oil and gas company, and disable radar systems. Additionally, DDoS and web page defacements have been a frequent tactic among states. While China appears focused on key point strikes, they have also been developing a scattershot approach through hacker groups, militias, and online people's war. Combining these two approaches, along with conventional strikes, can create a synergy.

Computer network exploitation (CNE) encompasses the theft of data, intelligence gathering, embedded backdoors, and mapping systems for attack. The theft of intellectual property is particularly useful for China, as it can skip generations in research and development, boost economic might, and modernize the country and military. Understandably though, Beijing is facing increasing international condemnation for such alleged activity. Its alleged use of embedded backdoors has also alarmed the international community. Despite strong denials of such conduct, Chinese information warfare theorists have been discussing "virus-infected" and remotely activated "self-destructing" microchips as early as 1988 (Thomas 2004, 45). The following chapter, *The Development of Computer Network Exploitation, 1998–2014*, will expound on these topics.

The PRC is becoming increasingly reliant on the Internet, and consequently, creating new vulnerabilities. As of 2012, China had 538 million Internet users; and China's Twelfth Five Year Plan calls for increased integration of the Internet and computer networks into business, government, military, and society as a whole. This modernization includes advancements in electronic payments and mobile phones, the creation of wireless cities and intelligent power grids, and the convergence of all communication medium. Meanwhile, computer attack and exploitation incidents targeting China have numbered in the tens of thousands annually and are rapidly increasing. In 2013 alone, CNCERT reported approximately 24,000 web page defacements, 30,000 watering hole websites targeting China, 47,000 Chinese IP addresses compromised by RATs, and

700,000 mobile phone malware infections. This chapter has also revealed at least three instances of China's critical infrastructure being attacked by non-state actors and four organizations which serve as traffic hubs being targeted by the United States. Further, there are an estimated 120 states developing CNO capabilities amid what could be viewed as a cyber arms race. China's negative reputation concerning CNE allegations could make China a primary target for these developing capabilities, either directly or as a proxy to strike other states. Additionally, all of the asymmetric benefits that CNO provides China as a tool for attack and exploitation—including plausible deniability, relatively low cost and entry level, and relatively few consequences—also apply to those who wish to target China. Lastly, China is attempting to influence CNO-related international law to strengthen its position. Among these is a trend of imposing access to the trade secrets of companies wanting to conduct business in China. This allows the People's Republic to identify exploits, reverse engineer products, and repel foreign competition.

NOTES

1. The terms virus and worm are often used interchangeably in the media; however, they are different. A computer virus is a malevolent piece of code which attaches itself to other programs and instructions on a computer. Once these programs or instructions are run, the virus runs along with them, seeking out additional hosts where it can replicate and attach. In addition to spreading itself, viruses can be designed to perform a variety of harmful actions, such as deleting files or creating glitches. In 2006 and 2007, the Panda Burning Incense virus infected millions of computers in China. It replaced the icon of infected applications with an image of a panda bear holding three incense sticks in prayer, or in some cases, this image was flashed on the computer screen. The Panda Burning Incense virus also stole passwords and financial details from infected computers (Areddy 2010; Fletcher 2009; Guo, Gu, and Wu 2011; Henderson 2009; Lemon 2007). A worm is similar to a virus; it can replicate, spread, and perform a variety of harmful actions; however, a worm does not need to attach itself to other programs. Worms are self-contained programs, capable of continual action, without the need for a user to unwittingly start an infected program. For this reason, worms tend to spread more quickly than viruses (Taylor, Fritsch, and Liederbach 2015, 29–30).
2. Air-gapped networks are physically separated from, and share no connections with, outside networks, including the Internet (Amoroso 2011, 63–64).
3. A tunnelling protocol is a workaround which allows a user to access a typically incompatible network.
4. An individual with no formal training can cause significant damage to a state with only an off-the-shelf computer and an Internet connection. For example, in 2001 and 2002 Gary ("Solo") McKinnon probed U.S. Army, Air Force, Navy, Department of Defense, and NASA computers causing $700,000 worth of damage, taking down a network of 2,000 computers, accessing classified data, and deleting and re-writing files (Boyd 2008). He accomplished this on his own from his home in London using commercially available software and a dial-up connection. McKinnon claims he was searching for proof that the United States is hiding information about UFOs and an anti-gravity propulsion system (previously documented in Fritz 2008). However, there are also highly advanced skills in hacking. A state seeking to develop CNO as a

military domain would need to be strongly competitive and take a comprehensive approach.

5. A security exploit is a prepared application that takes advantage of a known weakness. It is a piece of software, data, or commands that utilize a bug, glitch, or vulnerability to cause an unintended or unanticipated behavior to occur on computer software, hardware, or electronic devices. This can allow the attacker to take control of the computer, permitting its use for other tactics (Fritz 2008).

6. A Northrop Grumman research report commissioned by the U.S.-China Economic and Security Review Commission disputes the blue army's identification as a cyber warfare unit, particularly its offensive capability (Krekel, Adams, and Bakos 2012). Although the same report states that the group is used for online confrontation exercises and training, which could help develop or enable an offensive ability. These statements appear to be a response to media reports which exaggerated China's acknowledgment of a modest and defensive unit, rather than a challenge to its stated purpose or significance.

7. Despite the creation of the United States Cyber Command (USCYBERCOM) in 2009, there were concerns that it would be overshadowed by the Department of Defence or National Security Agency. Further, USCYBERCOM focuses on the .mil domains, while the Department of Homeland Security deals with .gov domains. A primary target of CNO, the .com domains, is not comprehensively covered by either organization. Additionally, the debate over a streamlined structure is not over, and some have called for cyber warfare (including CNO) to be given its own military branch, despite overlapping interests among all military branches (see Conti and Surdu 2009; Lynn 2009; Monroe 2009).

8. Internet security refers to CNO and Information Operations (IO), while informationization typically refers to NCW. This leading group may have been addressing all three branches; however, given the context of official statements, it appears that the term "informationization" is being used differently here. Here it is referring to using computer networks to streamline and modernize government and civilian operations. The primary function of these changes does not appear militaristic in nature, and they would likely be Internet-connected, making it fit the definitions of CNO and IO. Meanwhile, the majority of references Chinese officials make in regards to informationization remain non-Internet connected and militaristic in function; therefore, NCW remains a valid synonym elsewhere. In 2015, news reports began referring to this same leading group by the title "Central Leading Group for Cyberspace Affairs," perhaps in an effort to clarify its function (see Creemers 2015; Rawlinson 2015; Tham 2015). The Office of the Central Leading Group for Cyberspace Affairs is a dual nameplate for the Cyberspace Administration of China (CAC), which itself was formerly named the State Internet Information Office (SIIO) (see Chang 2015; Cheung 2015; Cyberspace Administration of China launches official website 2015; Segal 2014). These more recent reports place the organization's role more exclusively in the IO branch of cyber warfare. This not only reflects a complex and changing organizational structure, but also the close relationship between computer network defense and IO. Defensive policies on domestic use implemented by this organization would be legal under Chinese law, and therefore would not meet the hacking criteria of CNO. Nevertheless, this organization remains relevant to the discussion of CNO organizational difficulties in this instance.

9. See chapter 2 for further detail on the history, and alleged government support, of hacker groups.

10. A Distributed Denial of Service (DDoS) attack occurs when thousands of computers attempt to access a website at the same time with the intent of overloading network servers. This is similar to thousands of humans attempting to use a building's exit at the same time. The entryway is a chokepoint and cannot handle such high volumes of traffic in a short period of time, rendering the website (or exit) inoperable.

11. Web page defacement is comparable to electronic graffiti, and it is integrated directly into the website design (not for example, the comment section). These are

typically images and text messages, but they can also include audio and video. They are commonly offensive and are meant to send a message, evoke emotions, or demoralize an opponent (Taylor, Fritsch, and Liederbach 2015, 29).

12. It is unlikely that China would use an ASAT to disrupt U.S. space-based assets. To disrupt U.S. satellite dominance would require a large-scale sky-clearing operation, because the United States has constellations of satellites with multiple redundancy. For example, the U.S. Global Positioning System (GPS) provides tactical communication and precision navigation, making it a desirable target—however, GPS maintains a constellation of 31 satellites in orbit, providing 6 satellites in view from any position on earth at any time. GPS only requires 4 satellites to operate at full capacity, and 3 for reduced accuracy. When one is destroyed, others can be maneuvered to fill holes in the network, and not all of these satellites are within striking range at any given time. This means a sky-clearing operation would take a significant amount of time, thereby revealing Beijing's intention, allowing the United States to retaliate and maneuver its other satellites out of harm's way. It would also cause international dispute due to the resulting space debris. Additionally, there is no guarantee an attempt would be successful, as each launch requires precise targeting, and China's ASAT has only been fully tested once. It is more likely China would attempt to knock out the corresponding relay stations on Earth by using a cyber attack. Chinese tacticians have focused on neutralizing the uplinks and downlinks of the space-based systems through diverse forms of cyber attack including simple DDoS attack. This gives the advantages of deniability and low cost. It would remove distance from the equation, allowing multiple targets to be taken out simultaneously regardless of location, and it would minimize international condemnation and/or involvement.

China could destroy a vast amount of U.S. electronics, including computers, cars, phones, and the power grid, using EMP weaponry. This is something of which all nuclear-armed states are capable by means of high-altitude nuclear explosions, taking as few as three to blanket the continental United States (Electromagnetic Pulse 2005). Open source materials have shown the United States, China, France, and Russia all using an EMP burst as a surprise first strike in war games (China's Proliferation Practices 2008; Liang and Xiangsui 1999; Winn 2008). However, it is unlikely China would use such brute-force tactics. Using a high-altitude atomic burst would cause international outrage as it damages the planetary environment, and it indiscriminately disrupts everything in its blast radius. Alternatively, shutting down the U.S. power grid, production lines, water utilities, telecommunications, or modes of transportation are all possible through cyber attack, and it would provide the benefit of increased anonymity.

13. A botnet is a large collection of computers that are under illicit remote control and can be banded together. Infected computers, known as bots or zombies, can be spread across the globe. The botnet controller, or bot herder, issues commands through a command and control (C2) server. This way, if a C2 server is detected and shut down, the bot herder can begin issuing commands from a different C2 server. The bot herder can also "drop" and store weapons payloads onto these servers for the bots to retrieve. End users are often unaware that their computer is being used as a bot (Amoroso 2011, 6–9).

14. Short for malicious software, malware is a blanket term for harmful hacking tools (Lawrence and Riley 2013; Taylor, Fritsch, and Liederbach 2015, 30).

15. Another individual, Jonathan ("c0mrade") James, committed suicide while under investigation for connection to the TJX hacker ring (Poulsen 2009). He had previously been charged with penetrating NASA computer networks, where he downloaded $1.7 million worth of software used to control the International Space Station's life support (Harrison 2000).

16. Some satellite providers offer Internet access as a for-sale service. However, this is not a requisite for satellites to be considered Internet-connected. Many satellite networks use Internet connectivity as a part of their operations, but they do not offer it as a customer service.

17. For a comprehensive examination of satellite hacking, including architecture design and timeline of known incidents, see Fritz 2013a.

18. A Trojan horse, or Trojan, is a self-contained program which masquerades as a beneficial program while secretly performing harmful functions. They can contain varying packages of malicious tools (Taylor, Fritsch, and Liederbach 2015, 29).

19. It is uncertain if these computers were connected to the Internet, or if the modern Internet existed at this time; however, it points toward a much earlier development stage of computer network attack than Stuxnet.

20. Spear phishing is a targeted e-mail scam which attempts to gain unauthorized access to a target's computer. The e-mails, which may appear to be legitimate and from a known source, contain a hyperlink or attachment which, if clicked on, will infect the computer. Common attachment file types used in spear phishing include Word and PowerPoint documents, PDF documents, and image files.

21. Governments are aware of the threat CNA poses to ICS, and they have been taking steps to increase personnel screening, inspections, inter-agency communication, emergency response capability, scrutiny of sensitive high-tech parts produced abroad, and overall computer network defense. ICS systems may be more robust than some reports have indicated. Moreover, these systems are designed to be distributed, diverse, redundant, and self-healing, in part because weather systems and natural disasters pose a continual threat of disruption. A cyber attack against ICS systems may require a sustained assault against multiple targets to have a significant effect. Additionally, it is not entirely automated as humans are retained for monitoring.

22. Television programs and Hollywood films, such as *Die Hard 4.0*, *Hackers*, and *War Games*, which were set in the present day, give the impression that an intruder can enter any network with a few simple key strokes and the phrase "I'm in." Further, they create the illusion that an intruder can jump between multiple databases and compile information instantaneously, as if banking details, building blueprints, facial recognition software, medical records, missile launch codes, mobile phone tracking, personal history, surveillance cameras, television and radio broadcasts, traffic lights, transportation systems, utility systems, and other elements are all operating on the same network with an easy-to-use navigation and interface. While each of these could theoretically be hacked, they represent a wide range of constantly changing and diverse systems. Even a state, with all of its resources, would be unlikely to accomplish such a feat in one city, let alone country wide, or globally. Focusing on critical nodes would be crucial, as would mapping of systems beforehand, cataloguing known exploits, and maintaining a pre-strike presence within networks. Smaller, less technologically advanced states might provide an easier target, yet at the same time, they would have less networks to attempt exploitation on. Such exaggerated portrayals can also have the opposite effect, causing policymakers and the public to dismiss a credible threat.

23. Striking hubs, critical links, or the largest suppliers of energy distribution could cause other systems in the network to collapse. For example, by knocking out the largest suppliers, other nodes will attempt to compensate for the loss and may in turn overload.

24. The People's Republic of China maintains that Taiwan belongs to it following the end of the Chinese Civil War in 1949, while the United States maintains it is obligated to act in Taiwan's defense under the Taiwan Relations Act of 1979.

25. This is often referred to as social engineering. It is a human interaction element to hacking, comparable to a con artist, and it is a prized trait within hacking culture (Abagnale 2000; Mitnick 2002; Raz 2013).

26. Zero-day exploits are exploits which are not yet known to the public or IT companies. Once an exploit is known, it might be patched (repaired or blocked) and no longer of use to the hacker. For this reason, hackers catalogue these vulnerabilities and keep them secret until they are ready to use or sell them.

27. There are many legitimate uses for remote access, such as network administrators helping users with computer problems; however, it can also be used by hackers to obtain unfettered access to a victim's computer from anywhere on the globe.

28. A payload is the portion of malware which performs damaging effects. The name can also refer to the type of damage it causes. A payload can be thought of as the arsenal, or weapons load, of malware. It is sometimes used interchangeably with the terms toolkit and rootkit, although a rootkit requires administrator access and is particularly hidden, or buried, within the system.

29. A keylogger is software which records all of the keystrokes made on the victim's computer. Among the information this can yield are user passwords. Tangible, hardware-based, keyloggers can also be covertly attached to concealed USB ports or cable connections for later retrieval.

30. Using Taiwan as a proxy to attack U.S. targets has some added benefits, such as creating friction between the United States and Taiwan, and causing U.S. officials to choose their words carefully when assessing the situation given the One-China policy. For example, stating that China was responsible for a CNE incident (routed through Taiwan) could be construed as an admission that Taiwan is a part of China.

31. An IP address is a numerical identification assigned to devices participating in a computer network utilizing the Internet Protocol (TCP/IP) for communication between nodes. In essence, each computer has its own unique IP address.

32. Vulnerability scanners may be used to identify known weaknesses. One such scanner known as a port scanner automates the process of finding weaknesses of computers on a network. These check to see which ports on a specified computer are "open," available to access, and sometimes will detect what program or service is listening on that port (Fritz 2008). Further, port scanners can identify Trojan horses installed on a computer by other individuals, which allow remote access. This allows the intruder to use the backdoor already installed by another unknown intruder (Taylor, Fritsch, and Liederbach 2015, 156).

33. A watering hole attack is a technique whereby the assailant booby-traps a website which is popular among a specific group. For example, if the assailant wants access to a specific organization's network, they can attempt to determine which websites are popular among employees, or possibly a website which is regularly used to aid their work. The website, or web page, is then altered to infect visiting IP addresses, and the assailant waits for its targets to visit the corrupted website on their own "like a lion will lie in wait to ambush prey at a watering hole" (Gragido 2012). This can achieve the same effect as spear phishing, without needing to trick a user into clicking on a harmful link.

34. Soft power is the ability to obtain foreign support through appealing culture, policies, and values, rather than through the use of force, payments, or threats (Nye 2004).

35. "Spam mail" is listed as the second most reported type of incident in 2007, which brings into question CNCERT's definition of the word "incident." This could be due to CNCERT difficulties in providing an English translation.

36. The 2013 CNCERT/CC Annual Report also reports 11.4 million IP addresses in mainland China as being "infected with [a] Trojan or [belonging to a] Botnet." This might be an adjusted cumulative total, rather than an annual increase. However, the language used is unclear, and it appears to contradict figures given elsewhere in the report. For example, being a part of a botnet implies malicious remote control, and being infected with a Trojan could result in malicious remote control, yet only 47,300 IP addresses are listed as such. It is possible these computers were infected, but not utilized. CNCERT's 2010 and 2007 annual reports list 15.9 million and 4.6 million Trojan or botnet infections in their respective years.

37. The Aum Shinrikyo cult, responsible for the 1995 Sarin gas attacks on the Tokyo subway system, was discovered in 2000 to have also developed a software program capable of tracking police vehicles. Further, the cult had a role in the development of software used by "80 Japanese firms and 10 government agencies" leading to concerns

that they may have installed backdoors to facilitate cyber attacks (DCSINT 2005). In addition to showing what CNA is capable of, this is also a cause for concern for China, who must also defend against cults and terrorist activity, such as the Falun Gong or the 2014 Kunming railway station attack in which 31 people were killed and 141 were injured (Yamei 2014).

38. A logic bomb is a harmful computer application which will activate if set conditions are met. For example, deleting an employee's name from a payroll database (signifying their termination) could be set as the trigger to activate the harmful application.

39. U.S. threat perceptions of China, also referred to as the "China threat theory" by PRC scholars, has attracted a wide academic literature (Broomfield 2003; Dellios and Ferguson 2013; Tiezzi 2014).

40. Events which fuelled concern over U.S. hegemony in the early twenty-first century included wars in Iraq and Afghanistan, aggressive policies under the Bush administration (2001–2009), the increased use of unmanned combat aerial vehicles, and the Obama administration's (2009–2017) rebalancing, or pivot, toward the Asia-Pacific (Channer 2014; Clinton 2011; Miller 2011; USCC Annual Report 2014).

41. Russia's use of CNO predates China and includes alleged involvement in the 2007 cyber attacks against Estonia and the 2008 cyber attacks surrounding the Russo-Georgian War (Fritz 2008). Russian military intervention in the Ukraine in 2014, including the annexation of Crimea and alleged support of separatist fighters who shot down Malaysia Airlines Flight 17, coupled with an increased military presence which includes renewed bomber runs, has led some analysts to question if Russian relations are returning to a Cold War status (Leftly 2014; Pollard 2014; Williams 2014).

42. The Distinguished Warfare Medal was placed under review and production was halted; however, the concept continues to be developed, and the prominence of cyber and drone warfare in U.S. strategy has not diminished.

43. This type of attack is known as DNS cache poisoning. Incorrect data is illegally entered into a server's Domain Name System (DNS) cache, so that it returns the wrong address for queries.

44. This was a high-profile incident, known as Operation Aurora, and it is detailed in chapter 2.

TWO

The Development of Computer Network Exploitation, 1998–2014

This chapter will examine the development of Chinese computer network exploitation (CNE) ranging from individual incidents, recurrent targets, and advanced persistent threats (APTs) to hacker groups and state-sponsored groups. Incidents will primarily be examined in order of occurrence, or for APT, the earliest known activity. This does not always correspond to the date the information was released. For instance, Mandiant's report, *APT1: Exposing One of China's Cyber Espionage Units*, was released in 2013, but the CNE campaign it revealed began in 2006, so it is discussed in the 2002–2006 section. Additionally, supporting information is used throughout to aid in the formation of cohesive stories, which will reveal more about Chinese CNE than a strict timeline could. Therefore, this chapter is largely, but not entirely, chronological in organization. The period from 1998 to 2002 was dominated by the formation of hacker groups and included a number of large-scale cyber conflicts. These groups set the foundation for the widespread CNE which followed from 2000 to 2014. In total, this chapter details more than 100 alleged Chinese CNE incidents and 19 APT which targeted 71 countries (see appendix A).

Significant APT, or campaigns, examined in this chapter include Nortel (2000–2009), Byzantine (2002–2013), Titan Rain (2002–2006), NetTraveler (2004–2013), Avocado (2005–2008), APT1 (2006–2013), Shady Rat (2006–2011), Sykipot (2006–2013), Axiom (2008–2014), Night Dragon (2009), GhostNet (2009), Operation Aurora (2009), Hidden Lynx (2009–2013), the Shadow network (2010), Ke3chang (2010–2013), Nitro (2011), Luckycat (2011), Icefog (2011), and Lurid Downloader (2012).[1] The investigation into APT1 was unique in that it also revealed an alleged CNE group, PLA Unit 61398, including details of personnel and facilities, rather than broadly attributing the CNE to the Chinese government as a whole (Mandiant 2013). This unit has also been known by the names Comment Crew, Comment Group, Comment Panda, Shanghai Group, and by the name of another APT they conducted known as Shady Rat. A separate group operating out of Shanghai, known as Putter Panda or MSUploader, has been identified as PLA Unit 61486, although it has not been ascribed to one of the specific APT operations listed above. Another group conducting CNE for China is the Beijing Group, which has been determined to be synonymous with the Elderwood Gang and Sneaky Panda, and was responsible for Operation Aurora (Clayton 2012). A fourth unnamed group operating out of Chengdu, PLA Unit 78006, may be responsible for the Shadow network, or less clear, Byzantine (Stokes, Lin, and Hsiao 2011). These groups' identities are the most firmly established in open source material, yet they represent only four of the estimated 20 Chinese state-sponsored CNE groups (see appendix B for more) (Clayton 2012; Mandiant 2013; Riley and Dune 2012; Segal 2012). Beyond isolated incidents, APTs, and the groups conducting them, there are recurrent targets and themes, such as CNE targeting the National Aeronautics and Space Administration (NASA) or information on the F-35 aircraft, and embedded backdoors in products, also known as Manchurian Chips or Trojan Dragons. These are discussed as collections of incidents when appropriate; however, they are not as cohesive as a campaign or attributed to a specific group. Some defense contractors, government agencies, and corporations have been frequent targets; yet intrusions into their networks does not necessarily represent a single continued and coordinated effort, and they could be on the target list of multiple Chinese CNE groups.

Despite the media's focus on the later CNE incidents (within this chapter's timeline of 2000–2014) being attributed to China, such as penetrations into Google revealed in 2010, or the Mandiant report released in 2013, evidence suggests that China had been conducting sustained and widespread CNE throughout the decade 2000–2010. Allegations often stop short of directly stating the Chinese government or military was responsible, yet it is strongly implied and then typically followed by a caveat on the difficulties of attribution in cyberspace.[2] There are several reasons for this. Online anonymity, the ability to route traffic through multiple computers around the globe, and the ability to remotely control computers do pose attribution problems. Further, governments and companies do not want to damage relations with China, in large part due to the economic benefits of their relationship, yet they must also protect national security and intellectual property. There are also legal ramifications to accusations which must be considered, and revealing too much information on detection methods could place that capability's continued use at risk or allow other states to obtain an equal capability. The following collection of elements are often used to infer the role of the ruling Communist Party of China (CPC) in CNE.

(1) Motive in target selection. As an example, information relating to contract bids, government policy agendas, military operational procedures, political activist communications, or the seismic surveys of oil companies are unlikely targets for non-state actors. For hackers, they lack the attention-grabbing appeal of other targets, and for criminal organizations, they do not easily translate into profit. Hacktivists or terrorist organizations might have an interest in this data, yet it does not seem directly applicable to achieving their goals, and it is unlikely to be worth the resources and risk in comparison to other targets. As another example, highly technical information relating to engineering, science, and technology requires vetting that is beyond the capability of many non-state actors (Krekel, Adams, and Bakos 2012). This list is not comprehensive; the key point is that motive in target selection is one factor in inferring attribution. In the case of APT1, targets included "at least four of the seven strategic emerging industries that China identified in its 12th Five Year Plan" (Mandiant 2013). In some cases, the same group that is eavesdropping on Chinese dissidents will also target aerospace technology; a dual motive that points toward China (Perlroth 2013; USCC Annual Report 2009).

(2) Operational logistics. Operations which require a large amount of funding, resources, and staff point toward state sponsorship. Many of the incidents described in this chapter, particularly the APT, were large-scale undertakings involving hundreds of command and control (C2) computers and hundreds of targets. The ability to automate this collection process is limited in part by a need for continual monitoring and adaptability in order to avoid detection, maintain access, and locate key information. Development of malware could possibly be purchased, yet the spear phishing and watering hole attacks common in Chinese CNE often involve individually tailored social engineering. In some cases, Internet Protocol (IP) addresses and malware are also tailored to specific targets, taking place within the larger campaign (Novetta 2014). Combing through the large amounts of data collected in order to obtain benefit from it would also be time consuming. An APT targeting technical information across multiple industries could require experts from each of those industries, and possibly linguists, to know what to look for and to understand what is retrieved (USCC Annual Report 2009). Operations which run for multiple years also point toward the need for a large amount of funding and staff, and in some cases intruders have even been noted to be operating on a regular work schedule, weekdays from 9:00 a.m. to 5:00 p.m. Beijing time (Clayton 2012; Demick 2013; McAfee 2011).

(3) Final destination of stolen information. Digital trails which end in China provide an additional element of confidence when assigning attribution. Due to the ability to route traffic through multiple computers and remotely control computers, the identity of the person conducting CNE may not be known for certain. However, when CNE is traced back to IP

addresses and servers in China, this does not remove the possibility that they are in fact the last link in the chain. In some cases, these addresses can be further linked to PLA institutes, members, and units. In other cases, insiders are also a part of CNE operations. The arrest of Chinese nationals attempting to return to China with stolen property furthers the connection that the electronic component of their operation was also based in China.

(4) Registration and online discussion. Some of the domains and e-mail addresses used to conduct CNE can be traced back to the person who registered them. For example, an e-mail address used to register a domain or conduct spear phishing might also be carelessly used by the same individual elsewhere online, such as when websites require an e-mail address to post comments or create a profile. Searching for instances of dual registration has been used to reveal connections between CNE incidents and city locations, city records, hacker groups, malware development, phone numbers, resumes, state-sponsored groups, and university affiliations (CrowdStrike Intelligence Report 2014; Information Warfare Monitor 2010; Mandiant 2013). These can strengthen confidence in attributing the individual to malicious activity and strengthen confidence in attributing China as the final destination of stolen information. For example, an IP address used to conduct CNE and associated with someone who discusses malware development in blogs and forums seems less likely to be an unknowing relay station and more likely to be the perpetrator (Demick 2013; Henderson 2007b).

(5) Mandarin language. The use of Mandarin language within the programming and control of malware and C2 infrastructure is another indicator of possible Chinese attribution. For example, 97 percent of the "remote desktop sessions" observed in APT1 were conducted while using a Chinese (Simplified) keyboard layout (Mandiant 2013). Despite the ability for malicious code to be revealed online and subsequently used by any actor around the globe, the use of some malware families in itself remains a possible indicator of attribution. Due to language barriers, benefits of early access, and a need for developing malware indigenously, the deployment of some malware families remain largely contained to their place of origin (Henderson 2007a, 15). This could also be due to established operating procedure and cultural preferences of the groups involved.

(6) Visible use of stolen information. In some cases, which will be discussed later in this chapter, stolen information is visible in Chinese products. Examples include code from Cisco, CyberSitter, and Sinovel; automobile designs; and components of the F-35 and F-22 aircraft. These go beyond similarities in products to distinct artifacts, or watermarks, belonging to the original company. China, or Chinese companies, may not have sponsored the theft, yet they did come into its possession and are therefore providing support at least in an indirect way (USCC Annual

Report 2009). Less visible, yet a cause for suspicion, are insights which appear to have been gained through CNE prior to international government discussions and contract bids.

The incidents described within this chapter demonstrate extensive CNE with substantial impact across all major industries. A total of 71 countries are noted as having been the targets of Chinese CNE, however, the majority of these incidents are focused on the United States of America. This could be due to a higher rate of U.S. reporting, a higher number of desirable targets within the United States, and/or greater ease in accessing English-language documents. Even within the United States there is a perceived "persistent underreporting of events" (USCC Annual Report 2010). In addition to not wanting to damage relations with China, companies are reluctant to disclose intrusions because it could damage consumer and investor confidence, and personnel withhold information out of fear of being held responsible for a breach. The incidents in this chapter are also restricted to unclassified material. Given these limitations of language, underreporting, and open source material, it is possible that this only represents a small portion of all Chinese CNE conducted over the 17-year period of 1998–2014. The 2013 U.S. IP Report stated that the United States lost approximately $300 billion per year through the theft of intellectual property over computer networks with 50 to 80 percent of the total originating from China (USCC Annual Report 2013). The director of the National Security Agency (NSA) and commander of U.S. Cyber Command, General Keith Alexander, referred to it as "the greatest transfer of wealth in human history" (The IP Commission Report 2013). Despite U.S. CNE, such as the acknowledged PRISM surveillance program, Washington states that the PRC is unique in its targeting of commercial intellectual property rather than strictly intelligence and military targets. Further, CNE can be a stepping-stone for computer network attack (CNA), either by mapping systems and their weaknesses or by loitering within networks and setting up the capability to attack if it is needed. Therefore, computer networks which have been targeted by CNE, including commercial entities, could become CNA targets in a time of conflict.

RISE OF HACKER GROUPS, 1998–2002

Alongside an increase in People's Liberation Army (PLA) literature discussing computer network operations (CNO), and the rapid increase of Internet access within China, solidified hacker groups began to emerge around 1998. Between 1998 and 2002, these hacker groups were involved in large-scale cyber riots with Indonesia, Japan, Taiwan, and the United States, which were comparable to the more often cited, yet later occurring, Estonian cyber conflict of 2007. The hacker groups' activity during these conflicts was primarily CNA, not CNE. However, it forms the beginning of Chinese CNO and establishes the groundwork, including anonymity, capability, and talent pool, for Chinese CNE which became dominant around 2002. Some of these hackers have gone on to careers with the government or Information Technology security companies within China, and they have inspired a new generation to learn hacking skills. Further, these hacker groups continue on as a supplemental tool for state-sponsored CNE by providing malware development, plausible deniability for some actions, and serving as a militia. Chinese hackers appear distinctive in their strong nationalist sentiment, although to some degree "very strict internal guidelines" and "support of the government" had to be maintained to prevent them from being censored or shut down (Henderson 2007a, 10). They are not averse to seeking individual fame and profit, yet nationalism remains a strong characteristic overall.

In times of economic trouble, Southeast Asia's wealthy Chinese diaspora have historically been treated as a "scapegoat" (Chua 2004, 2–6, 23). Indonesia was no exception. The most recent instance of this occurred during the May 1998 "Riots of Indonesia." These riots grew out of economic and political dissatisfaction with the prevailing New Order regime of President Suharto and occurred in the immediate aftermath of the 1997 Asian Financial Crisis which was particularly damaging for Indonesia (Collins 2002; Franciska 2014). Out of the turmoil of 1998 emerged Indonesia's political reformation (*reformasi*), but not before rioters turned on the Chinese community, blaming it for monopolizing the wealth of the nation. The destruction of ethnic Chinese lives and property was a catalyst for bringing together "relatively independent cells" of hackers within China (Henderson 2007a, 16; Honker Union of China to launch network attacks against Japan is a rumor 2010). Some of the larger cells included China Eagle Union and The Green Army. They formed the Chinese Hacker Emergency Conference Center for communication and coordination and began linking to each other's web pages, forming a larger network. This loose connection of groups utilizing redundant communication and resources, including Internet Relay Chat (IRC), is comparable to the group Anonymous (see Fritz 2008, 35–37). Their actions against Indonesia included DoS attacks, e-mail bombs, password cracking, and web page defacements. Key events in 1999 which sparked additional cyber conflicts were the U.S. bombing of the Chinese embassy in Belgrade and Taiwanese president Lee Teng-hui's proposal for a state-to-state relationship with China (Krekel 2009; Honker Union of China 2010; Thomas 2004, 27–28; USCC Annual Report 2009). Two key hacker groups emerged in 2000, the Red Hacker Alliance, also known as the Honker Union of China, and Javaphile, both of which have connections to CNE incidents discussed later in this chapter (Henderson 2007a, 3–5). Cyber conflict with Taiwan continued in 2000 after the pro-independence candidate Chen Shui-bian was elected president, and a new cyber conflict with Japan began over a court case which represented a denial of the Nanjing massacre.[3] Hostilities with Japan continued in 2001 following a string of incidents perceived as slights to China: Japan's response to a traffic accident involving brake failure of a Mitsubishi automobile, poor treatment of Chinese passengers on a Japan Airlines flight, the approval of Japanese history textbooks which removed mention of World War II atrocities, and a visit by Japanese Prime Minister Junichiro Koizumi to the Yasukuni shrine in Tokyo that commemorates Japan's war dead, including convicted war criminals (Henderson 2007a, 38–41). These cyber attacks were not one-sided; Japanese and Taiwanese hackers also targeted Chinese websites.

Another China-U.S. cyber conflict erupted in 2001 following the collision of a U.S. reconnaissance/spy plane and a PLA Navy F-8 fighter jet, in which the Chinese pilot died (Brenner 2005; Thomas 2004, 55; USCC Annual Report 2009). This event is known as the Hainan Island incident, or EP-3 incident; although these titles are not always in reference to the CNA which occurred as a result. Patriotic Chinese hackers defaced dozens of U.S. military and computer industry websites, as well as the White House website. Patriotic U.S. hackers responded with inflammatory web page defacements, comment spamming, and the posting of photoshopped pictures. In what may represent an escalation and change of tactics, the EP-3 incident coincided with the release of the Code Red and Code Red II worms. At the time these were the most successful worms in Internet history, causing nearly two billion dollars in damages and infecting over 600,000 computers. The worms, which may have originated from a university in Guangdong, southern China, attacked computers running Microsoft's IIS web server and exploited a buffer overflow (United States General Accounting Office 2001). Home computers were largely unaffected; however, any attempt at infection caused them to crash. The worms created decreases in Internet speed, knocked websites and networks offline, and defaced websites with the phrase "Hacked by Chinese!"—although Chinese involvement was never confirmed. Code Red II had a slightly different payload that could open a backdoor, leaving the computers vulnerable to further exploitation, and may represent an additional shift in tactics from CNA to CNE (Rhodes 2001, Schwartz 2007; Cost of "Code Red" Rising 2001). These worms were followed by the release of the Code Blue worm, which was allegedly a response by U.S. hackers. It sought out systems infected by Code Red and reprogrammed them to launch attacks against targets in China. In particular, it launched DDoS attacks against the Chinese security firm NS Focus (Onley and Wait 2006; Delio 2001).

The EP-3 incident is also a key event, because it shows a rough method by which the party-state can control hacker activity, and in conjunction, its endorsement of hacker activity. Initially government officials were silent or encouraged hacker activity during the EP-3 incident. Later, as the cyber conflict escalated and a diplomatic solution was sought for the incident's non-cyber aspects, the CPC released statements through the People's Daily, the Beijing Broadcasting Institute, and others, including direct statements through government organizations such as the Public Security Ministry's Internet Safety Bureau stating that the actions of Chinese hackers were illegal (Henderson 2007a; China's Proliferation Practices 2008; Krekel 2009). In addition, text messages, e-mails, and phone calls were placed to influential hacker group leaders who then passed the message to stop down to their members. The PLA's connection to hacker groups and hacking incidents during this period could be greater than is commonly stated. A documentary on cyber security broadcast on China Central Television 7 (CCTV-7), the government's military and agriculture channel, showed what appeared to be PLA officers using point-and-click software to conduct a DDoS attack "against a Falun Gong–related website hosted on a network at the University of Alabama" in the United States (USCC Annual Report 2011). The footage shown is believed to have taken place in 2001.

Attacks by Chinese hacker groups have continued. In 2014, hundreds of Vietnamese websites were knocked offline or defaced following disputes over sovereignty in the South China Sea (Chinese hackers attack 745 Vietnam websites in a week: report 2014; Martin 2011). China is estimated to have 250 hacker groups "that are tolerated and may even be encouraged by the government" as an aggressive non-state actor that supports the government's agenda and provides plausible deniability (USCC Annual Report 2008; China's Proliferation Practices 2008). The utility of Chinese hacker groups may be further enhanced in a time of conflict when they could serve as a reserve militia and engage in online people's war. However, what is important to the development of CNE is that this serves as the beginning of China's entrance into CNO. Some of these early hackers have progressed to careers in information security and the PLA. As hackers begin to age and complete university degrees, the promise of a stable and legal career that accords with their interests and ideology provides an incentive for seeking government employment. The first isolated incident of alleged Chinese state-sponsored CNE occurred in 1999, when intruders seeking information on U.S. nuclear weapons downloaded data from the Los Alamos National Laboratory (Gertz 2000). A significant shift from quasi-endorsed hacker groups to a focus on alleged state-sponsored CNE conducted by fully trained and employed soldiers occurred around 2002. The first open source Chinese APT was a series of coordinated intrusions between 2000 and 2009 that targeted the Canadian-headquartered multinational telecommunications firm Nortel Networks Limited, which later filed for bankruptcy. Using stolen passwords, the intruders "downloaded technical papers, research-and-development reports, business plans, employee e-mails and other documents" (Gorman 2012, Nortel hit by suspected Chinese cyberattacks for a decade 2012).

EMERGENCE OF EXPLOITATION, 2002–2006

The first widely reported CNE campaign,[4] believed to be conducted on behalf of the CPC, occurred roughly between 2002 and 2006 and is known by the U.S. code name Titan Rain (Grow and Hosenball 2011; USCC Annual Report 2008). Sources vary on the start and end dates of Titan Rain, and multiple reports of computer intrusions during this period are not labelled as Titan Rain yet they fit the profile of this coordinated effort. Titan Rain targeted U.S. aerospace, defense, and government installations, including the Army Information Systems Engineering Command, Boeing, the Bureau of Industry and Security, the Department of Defense, the Department of Homeland Security, Lockheed Martin, the Missile Defense Agency, NASA, the National Nuclear Security Administration, the Naval Ocean Systems Center, Northrop Grumman, Raytheon, Redstone Arsenal, Sandia National Laboratories, and the State Department (Almeida 2006; China's Proliferation Practices 2008; Leyden 2007; Tkacik 2007). The information gathered included "aerospace documents with hundreds of detailed schematics about propulsion systems, solar panelling and fuel tanks for the Mars Reconnaissance Orbiter," specifications for the "aviation-mission-planning system for Army helicopters," plus "Falconview 3.2, the flight-planning software used by the Army and Air Force" (Thornburgh 2005). Among the alleged Chinese intrusions which might be a part of Titan Rain are "294 successful hackings into the U.S. Department of Defense computers" in 2003 and penetration of "about 150 Homeland Security Department computers" in 2006 (Committee on Foreign Affairs 2011). Unsurprisingly, penetrations into U.S. government networks in 2006, and concern over embedded backdoors, meant that the State Department would no longer source its computers from the Chinese computer manufacturer Lenovo (Tkacik 2008). In some cases, such as the Department of Commerce's Bureau of Industry and Security, entire networks had to be taken offline and the computers needed to be replaced in order to ensure their integrity (USCC Annual Report 2006). China is said to have "downloaded 10 to 20 terabytes of data," possibly from the Nonsecure Internet Protocol Router Network (NIPRNet) alone, and beyond theft of data, there was concern that China could alter data on networks (Krekel 2009; Tkacik 2008; USCC Annual Report 2008). The United Kingdom also reported penetration of computer networks under the Titan Rain campaign (Espiner 2005; Norton-Taylor 2007).

Incidents which are less likely to have been a part of the Titan Rain campaign include the use of repeated zero-day exploits against the U.S.

Department of Defense. These intrusions were traced to the Chinese hacker group Network Crack Program Hacker (NCPH), who were reported to have indirect ties with and occasional sponsorship from the PLA (Saporito and Lewis 2013). However, NCPH was operating out of Chengdu, central China. While this could correspond to PLA Unit 78006, Titan Rain operators were reported to be based in the southern city of Guangzhou (Stokes, Lin, and Hsiao 2011; Tkacik 2008). Any connection would likely have been a supplementary one, possibly in malware development. In another incident, the National Journal alleged that the Northeast blackout of 2003, which affected 55 million customers in the United States and Canada, was caused by Chinese intruders (Committee on Foreign Affairs 2011; Harris 2008). This contradicts official U.S. statements that the blackout was caused by a software bug. Further, if the National Journal's account is accepted as true, it is unclear if this was intentional CNA or CNE that had an unintentional consequence. Despite being an unverified incident, it is notable because it added to the China Threat perception and is an early example of concern over critical infrastructure that became more prevalent in later years. Additionally, in 2003, the Huawei Shenzhen Technology Company (Huawei) was accused of "stealing corporate secrets from U.S. counterpart Cisco Systems and wholesale pirating of Cisco's software" (Tkacik 2008; USCC Annual Report 2012). Over the next decade Huawei would be accused on multiple occasions of working on behalf of China's CNE activities.

Separate from Titan Rain, an APT campaign known as NetTraveler began in 2004 and continued through 2013. With over "350 high-profile victims in 40 countries," it targeted defense contractors, embassies, government institutions, oil industry companies, private companies, scientific research institutes, Tibetan and Uyghur activists, and universities (Kaspersky Lab 2013a). The year 2004 also marked the release of a new, allegedly Chinese, worm. The Myfip worm stole pdf files, with later variants targeting Microsoft Word documents, schematics, and circuit board layouts. Among the victims were "Bank of America, BJ's Wholesale Club, and Lexis-Nexis" (Brenner 2005). The worm not only stole intellectual property, such as product designs, but also took customer lists and databases. In the same year, "Taiwan's Ministry of Finance, the Kuomintang Party, the Democratic Progressive Party (DPP) and the Ministry of National Defense's (MND) Military News Agency" were subjected to network intrusions (Taiwan opposition party accuses China of hacking 2011). This was followed in 2005 with the Taiwan National Security Council being targeted by spear phishing that attempted to install backdoors in their network. Huawei also made news again in 2005 after India turned down "a planned $60 million" infrastructure investment due to "intelligence agency concerns" (Tkacik 2008). A new development in Chinese CNE appeared in the following year, 2005, and would be repeated in subsequent years. This was the combination of CNE and human espionage, or insiders. Having a physical presence at the target location can help bridge some of the gaps in CNE, such as aiding social engineering, locating targets and entry points, or causing the initial breach, like installing a backdoor or turning off a security feature, which can then be exploited further remotely. Examples from Belgium, France, and Sweden revealed the targeting of "various levels of European industry," "car-parts maker Valeo," and "unpublished and unpatented research," respectively (Luard 2005).

Between approximately 2005 and 2008, a sustained CNE campaign that originated from China targeted the computer systems of NASA. These intrusions and attempts to halt them were given the U.S. code name Avocado. Based on open source material, it is difficult to give an exact time frame for Avocado. The Titan Rain intrusions included NASA, yet articles discussing Avocado include incidents which occurred during Titan Rain's time frame, such as intrusions into the Marshall Space Flight Center in 2002 and Ames Research Center in 2004 where a technician was forced to disconnect "the facility's supercomputers to limit the loss of secure data" (Elgin and Epstein 2008). Further, alleged intrusions by China into NASA's computer networks did not cease in 2008. One possibility is that classified material is able to assign specific PLA units who were behind the intrusions and hence distinguish between intrusions based on the intruder rather than the date and target. Other possibilities are differences in exploitation tools and targeted data, which are not detailed in open source material. Regardless, the code name Avocado continually appears, and the allegation of intrusions under that code name remain. Within this time frame significant data breaches occurred at multiple NASA facilities including the Goddard Space Flight Center, Johnson Space Center, and Kennedy Space Center. Methods for gaining access included the familiar socially engineered spear phishing targeting key personnel. Data stolen included "budget and financial information" and "operational details of the Space Shuttle including performance and engine data"; however, the duration and extent of the breach suggest the losses may be greater and more specific than revealed (Elgin and Epstein 2008; Krekel 2009). Attempts to restore the integrity of compromised networks included the replacement of some computers.

Alleged Chinese CNE in 2006 continued to target the U.S. government, defense institutes, Taiwan, and commercial intellectual property. "Eight [U.S.] Congressmen and seven congressional committees," including "a vocal critic of China's human rights record" had their computer systems compromised (USCC Annual Report 2009; Krekel 2009). Intrusions into the networks of Fort Hood, National Defense University, and the Naval War College caused "$20 to $30 million in damage to each system," loss of data, and interruption of services including websites and e-mail (Krekel 2009; Committee on Foreign Affairs 2011). Taiwan also continued to be targeted, with the Ministry of National Defense (MND) and the American Institute in Taiwan (AIT) being compromised by spear phishing. "Account login credentials" were also stolen "from Chunghwa Telecom's Web mail system, the MND's telecommunications provider" (Krekel 2009). In addition to the theft of data, MND and AIT networks were used to spread misinformation. In total, 13 zero-day exploits "within Microsoft applications" were used against Taiwan in 2006, and the Taiwan coast guard administration discovered a hidden program that was sending shipping schedules "to an e-mail address in China" (Tkacik 2008). The combined use of CNE and insiders also continued in 2006 with the theft and attempted transfer of "4,000 Ford documents" to a "Chinese automotive company" (ONCIX 2011).

DEDICATED CNE GROUPS, 2006–2010

A series of APTs attributed to PLA Unit 61398 began around 2006. PLA Unit 61398 has at times been referred to as APT1, Byzantine Anchor, Byzantine Candor, Byzantine Foothold, Byzantine Hades, the Comment Crew, the Comment Group, Comment Panda, Shady Rat, and the Shanghai Group. One reason for these multiple names is that there is not always a distinction between an APT's title and the group conducting it. One CNE group might be responsible for multiple APT. Additionally, multiple names are the result of a lack of transparency (reliance on open source material) and a lack of coordination between those investigating incidents. For example, the intelligence community may use different names than IT security firms, and an organization discovering an APT may not be able to make a connection to earlier incidents, thus believing it is a new campaign or a new group warranting a new title (Clayton 2012). Further, the Byzantine monikers appear to refer to different phases of one APT, and this collection of APT phase titles double as a group title. Of all of the names associated with PLA Unit 61398, the Byzantine monikers are the least frequently cited as being connected to the rest (Lee 2013; Riley and Dune 2012). If Byzantine's inclusion was more clearly verified this group's activities could be changed to have a 2002 starting date in accordance with U.S. diplomatic cables leaked by WikiLeaks (Grow and Hosenball 2011). Additional conflicting information furthers justification for their separation. For example, some reports connect Byzantine to the GhostNet and the Shadow network APTs (discussed later in this chapter), with possible connections to the hacker group Javaphile and a base of operations in Chengdu (Glanz and Markoff 2010). Chengdu correlates with PLA Unit 78006 and the Shadow network; however, it does not correlate with GhostNet, and the majority of sources clearly place Shanghai as the base of operations for all of these titles except Byzantine monikers (Stokes, Lin, and Hsiao 2011; Information Warfare Monitor 2010; Mandiant 2013). It is possible that CNE groups provide occasional assistance to one another in APT campaigns, which could result in this type of overlapping attribution. For these reasons, Byzantine will be explored in proximity to examination of PLA Unit 61398 within this chapter but it will be treated as a distinct group/APT.

Byzantine (2002–2013) used the familiar technique of spear phishing to trick computer users into clicking on malware laden hyperlinks or attachments which would then allow the intruders to begin burrowing into the computer systems. German intelligence officials stated that Byzantine was targeting everything from "military, economic, science and technology, commercial, diplomatic, and research and development" data, with an increase in activity "preceding events like the German government's meetings with the Chinese government" (Glanz and Markoff 2010). France also accused China of being behind Byzantine CNE which targeted government officials and included the ability to turn on webcams and microphones for eavesdropping (Grow and Hosenball 2011). The European Union Council was also infiltrated, possibly in the quest for information on the European financial crisis and the Greek government's debt crisis. Some additional targets included law firms dealing with cases that were against, or affected, China; and major oil companies' "seismic maps charting oil reserves" or information regarding drilling in PRC-claimed territory (Riley and Dune 2012).

The most conclusive open source document tying PLA Unit 61398 to a CNE campaign was the 2013 Mandiant report, *APT1: Exposing One of China's Cyber Espionage Units* (McGarry 2013a; USCC Annual Report 2013). This report directly states that APT1, the Comment Crew, and the Comment Group are PLA Unit 61398; it indirectly states that Comment Panda and the Shanghai group are as well, and it states that Shady Rat is possibly connected. It is alleged that since 2006 this group stole "hundreds of terabytes from 141 organizations" located in 15 countries "spanning 20 major industries," including aerospace, energy, engineering services, financial services, information technology, high-tech electronics, satellites and telecommunication, and transportation (Mandiant 2013). Through these intrusions, the group gained access to "broad categories of intellectual property, including technology blueprints, proprietary manufacturing processes, test results, business plans, pricing documents, partnership agreements, and e-mails and contact lists from victim organizations' leadership" (Mandiant 2013). While many of the command and control servers were shut down or blocked after the release of Mandiant's 2013 report, PLA Unit 61398 is likely to be continuing CNE operations. Multiple sources corroborate Mandiant's assertion that Operation Shady Rat might have been conducted by the same group responsible for APT1 (Alperovitch 2011; Riley and Dune 2012; USCC Annual Report 2011). Operation Shady Rat was an APT, uncovered by computer software security company McAfee, which occurred throughout the period 2006 to 2011. It involved intrusions into at least 71 organizations located in 13 countries, which included businesses, defense contractors, government agencies, the ASEAN Secretariat, the International Olympic Committee, and the United Nations. Separately, intrusions seeking information on the presidential campaigns in 2008 of both Barack Obama and John McCain may also be linked to PLA Unit 61398 (Riley and Dune 2012). It is probable that other incidents noted below also belong to this group, or one of the other estimated 20 state-sponsored CNE groups. There is no open source evidence, however, to support a connection; so they will be treated as isolated state-sponsored incidents or new APT.

The year 2006 was also the beginning of a sustained campaign, or malware family, called Sykipot (Dutcher 2013). Like Titan Rain and Avocado, it is not attributable to a specific PLA unit. From 2006 to 2013 Sykipot "targeted defense contractors, telecommunications firms, computer-hardware makers, chemical companies, and energy utilities, as well as government agencies" in the UK and United States (Lemos 2011). Later exploitation attempts attributed to Sykipot include targeting the smart card identification used by the U.S. Department of Defense and attempting to "gather intelligence about the [U.S.] civil aviation sector" (Constantin 2012).

Similar to NASA and the Avocado campaign, there are some targets which are of continual interest to Chinese CNE. The Joint Strike Fighter program, or F-35 Lighting II, is one such target, and the first occurrence noted here was in 2007. While it has not been assigned a code name or APT title and it could represent sustained yet isolated incidents, the focus given to it by Chinese CNE warrants placing these incidents together for clarity. The approximately 900 subcontractors of the F-35 fighter program have been frequent targets of Chinese CNE; however, the incidents noted here are specifically referring to the F-35 as a target (USCC Annual Report 2012). In 2007 and 2008, intruders were able "to successfully exfiltrate several terabytes of data related to [F-35] design and electronics systems" (USCC Annual Report 2009). In 2009, "six to eight F-35 subcontractors were totally compromised," and between 2009 and 2011 the aircraft's manufacturers—Lockheed Martin in conjunction with Northrop Grumman and BAE Systems—were repeatedly targeted (USCC Annual Report 2012). Additionally, in 2009, eavesdropping on online conference calls allowed intruders to listen in on meetings where F-35 systems were being discussed, and in particular the "secure communications and antenna systems" may have been stolen (Reed 2012a). A series of weapons systems that were compromised in 2013 included the F-35, and the F-35 was one of the primary targets detailed in the 2014 indictment of Su Bin, both of which will be detailed further in their respective time frame below (Martinez, Meek, Ross, and Ferran 2013; Su Bin Complaint of Hacking for China 2014; Su Bin Indictment for Hacking for China 2014; US Attorney's Office 2014). Some officials have claimed that the secrets stolen by China were sensitive but not classified, and that a full blueprint of the aircraft has not been obtained by China. However; even small components can reveal weaknesses or cause delays and cost increases in F-35 deployment. Further, photographs appearing on Chinese military forums suggest that some of these systems have been integrated into the development of China's fifth-generation fighter aircraft prototype, the Chengdu J-20. Some analysts have speculated that photographs of an upgraded version of the J-20 in 2014 reveal a "new electro-optical targeting system under its nose" derived from the F-35, along with concealed engine nozzles and a different "radar absorbing coating" (Gertz 2014). These assessments are comparable to photographs revealed in 2012 which draw conclusions based on similarities between the cockpit and heads up display design of the J-20 and the US F-22 Raptor (Gertz 2014).

U.S. government agencies and defense contractors continued to be targeted in 2007. Intrusions into the Office of the U.S. Secretary of Defense forced a network of 1,500 computers to be temporarily taken offline (Committee on Foreign Affairs 2011; Krekel 2009; USCC Annual Report 2009). The U.S. State Department suffered "large-scale network break-ins," the Department of Homeland Security had 150 computers infected, and the White House convened a task force to determine if the use of mobile devices such as the Blackberry needed to be restricted within the administration due to CNE concerns (Tkacik 2008). A further undisclosed incident spurred a U.S. investigation which revealed a brand of computers produced by China's Lenovo company contained "beaconing activity" (USCC Annual Report 2012). Continuing on from Titan Rain, there were new allegations that China had penetrated the U.S. NIPRNet, including the development of software to disable NIPRNet during a time of conflict. One 2007 estimate suggested that Chinese hackers acquired 750,000 zombie computers in the United States alone; these could be used to conduct a DoS attack or act as relay and C2 stations for CNE (Waterman 2007). Chinese spear phishing continued throughout this year, targeting over 1,000 people employed at the Oak Ridge National Laboratory for science and technology, and Huawei again came under suspicion when it attempted to purchase 3com, a U.S. company "which supplies the U.S. government with security software, routers, and servers" (Krekel 2009; Tkacik 2008). The U.S.-China Economic and Security Review Commission concluded that in 2007, "the 10 most prominent U.S. defense contractors, including Raytheon, Lockheed Martin, Boeing, and Northrop Grumman, were victims of cyber espionage" (USCC Annual Report 2008). Moreover, an alleged Chinese state-sponsored CNE group known as Putter Panda, or MSUpdater, began operations in 2007 targeting the satellite and aerospace industries. Security technology company CrowdStrike has linked this group to PLA Unit 61486 (CrowdStrike Intelligence Report 2014; Novetta 2014). Chinese CNE had become so frequent and aggressive that U.S. President George W. Bush raised the subject to Chinese President Hu Jintao at the APEC summit in 2007 (Reid 2007).

On separate occasions China was publicly accused of hacking into government facilities by officials in Canada, Germany, Japan, New Zealand, and the UK. China was named as the primary concern for Canada's counterintelligence program, while the German Chancellery, Ministry of Economics and Technology, and Ministry for Education and Research were reported to have been infected by Trojans hidden in "Microsoft Word and PowerPoint documents" (Krekel 2009; Tkacik 2008). German officials stated that 60 percent of CNE targeting Germany originated in China. A separate survey of 625 manufacturers in Japan conducted by the Ministry of Economy, Trade, and Industry also concluded that approximately 60 percent of CNE originated from China, with 35 percent of those surveyed having "reported some form of technology loss" (ONCIX 2011). In the same year, the New Zealand Security Intelligence Service cautiously blamed China for intrusions into its country's computer networks (Tung 2007). Furthermore, UK government agencies were targeted, including the Foreign and Commonwealth Office, and the House of Commons was shut down after intrusions which apparently sought information related to Chinese human rights issues (Norton-Taylor 2007). MI5, the UK's domestic counterintelligence and security agency, issued "a confidential alert to 300 chief executives, accountants, legal firms and security chiefs warning of cyber attacks and electronic espionage sponsored by Chinese state organizations" (Krekel 2009).

Foreign allegations of Chinese CNE continued in 2008 with Australia, Belgium, Canada, France, India, and South Korea joining the list. India attracted a near daily occurrence of Chinese CNE targeting its government agencies, including the Ministry of External Affairs, as well as the private sector (Krekel 2009; U.S. Department of Defense 2009). South Korea reported losses of $82 billion in 2008 due to "foreign economic espionage" with 50 percent attributed to China (ONCIX 2011). Blurring the line between CNA and CNE, French embassies in Beijing, London, Ottawa, and Washington suffered web page defacements following French President Nicolas Sarkozy's meeting with the Dalai Lama (Krekel 2009). This also obscures the distinction between hacker groups within China and state sponsorship. The U.S. Department of Defense, defense contractors, think tanks, and the White House continued to be infiltrated by Chinese CNE in 2008 (Krekel 2009; Waterman 2008). As noted in 2003, the National Journal alleged two instances where Chinese CNE or CNA caused a U.S. blackout. The second instance was a blackout in Florida in 2008 that affected three million customers (Harris 2008; USCC Annual Report 2009). The National Journal report contradicts various other explanations reported as the cause of the blackout and attributes it to being accidentally triggered by Chinese CNE attempting to map the system.

The United States Transportation Command (USTRANSCOM) became the target of an unnamed APT campaign in 2008, or possibly it has simply been a target in multiple CNE incidents, continuing through 2014

(Mulvenon 1999). TRANSCOM provides air, land, and sea transportation for the Department of Defense in times of war and peace, and acts as a transportation supplement to the U.S. military. Intrusions into TRANS-COM networks, and those of its subcontractors, between 2008 and 2011 stole credentials, documents, e-mails, encryption passwords, flight details, user passwords, and computer code (Committee on Armed Services United States Senate 2014). In 2012, a commercial ship contracted by TRANSCOM had multiple on-board systems compromised, and a year-long investigation that concluded in 2013 revealed 20 successful intrusions into TRANSCOM attributed to China (Committee on Armed Services United States Senate 2014).

Additional CNE concerns raised in 2008 include an investigation into the possible theft of information contained on the laptop of U.S. Secretary of Commerce Carlos Gutierrez during an official visit to China, and "a leaked FBI briefing" which revealed concerns that China may have been embedding backdoors in Chinese-made "uncontrolled or counterfeit CSISCO computer routers" commonly used by "classified U.S. government and military computers" (Committee on Foreign Affairs 2011; U.S. Department of Defense 2009). This coincided with a series of incidents during the same year which relate to backdoor and supply chain security. In addition to the State Department ban in 2006 and the beaconing activity noted in 2007, Lenovo was accused of shipping software that contained malware in 2008. Furthermore, a digital picture frame manufactured in China was discovered to contain malware that could infect a computer when it was plugged into a USB port to receive pictures. Point-of-sale software or hardware manufactured in China and used in European retail stores was also discovered in 2008 to be transmitting the account and PIN information of transactions, using mobile phone networks, to "a suspected criminal syndicate with operations in Pakistan" (USCC Annual Report 2012). Seeking profit from individual bank accounts and the destination of the stolen data do not fit the profile of Chinese CNE; however, it does raise concerns over the integrity of Chinese supply chains and the capability of expanding operations.

A new Chinese state-sponsored APT named Axiom began in 2008. This APT was revealed in 2014 by a consortium of IT security companies including Cisco, FireEye, F-Secure, Microsoft, and Symantec. Axiom caused over 43,000 infections and was conducting CNE against a wide range of targets, including "Fortune 500 companies, journalists, environmental groups, pro-democracy groups, software companies, academic institutions, and government agencies worldwide" (Novetta 2014). Of these infections, 180 contained Axiom's top-tier suite of malware, suggesting that these were high-value targets or possessed higher levels of defense. A 2009 APT known as Operation Aurora, sometimes referred to as Hydraq after a Trojan horse it used, may have been conducted by the same CNE group that was responsible for Axiom. Google elevated the

Operation Aurora incident to high-profile status by issuing public statements, closing its Beijing office, and relocating its Chinese operations to
Hong Kong. Google stated that this move was a result of no longer being
willing to censor Internet search results; however, it was more likely the
result of the CNE conducted under Operation Aurora. Spear phishing
was used as one means of gaining initial access, and the intruders used
information found on social media, such as the Facebook, LinkedIn, and
Twitter accounts of key employees, to craft their messages (HBGary 2010;
USCC Annual Report 2010). Google reportedly had source code stolen,
and human rights activists who used Google's Gmail e-mail service were
targeted. It was later revealed that the intruders were also searching for
the names of Chinese spies, of whom the U.S. government may have been
aware, and who Google may have been tracking on behalf of the U.S.
government (USCC Annual Report 2013). While Operation Aurora is
commonly associated with Google, there were more than 20 companies
targeted, including Adobe Systems, Dow Chemical, Juniper Networks,
Morgan Stanley, Northrup Grumman, Rackspace, Symantec, and Yahoo
(Baldor 2013; HBGary 2010; USCC Annual Report 2010).

The CNE group which may be responsible for both Axiom and Operation Aurora has also been known by the names the Beijing Group, Elderwood Gang, and Sneaky Panda (O'Gorman and McDonald 2012). This
group is believed to be one of China's top two CNE groups alongside
PLA Unit 61398, and is by some accounts more sophisticated than PLA
Unit 61398 (Lawrence and Riley 2013; Novetta 2014; Stokes and Hsiao
2012). The Beijing Group may also be synonymous with a group that has
been operating from at least 2009–2013, known as Hidden Lynx (Doherty,
Gegeny, Spasojevic, and Baltazar 2013; Novetta 2014). Hidden Lynx has
been responsible for a number of CNE incidents and possible campaigns
including the targeting of Bit9, an information security company, and a
watering hole operation called VOHO that infected approximately 4,000
computers (Novetta 2014). This group does not target a single sector,
seeking out a wide range of commercial, defense, and government information, however they have displayed a particular focus on financial services. Adding further to the list of synonyms, Operations DeputyDog
and Ephemeral Hydra which will be noted in 2013, have been linked to
this same Beijing-based CNE group.

At least two other Chinese CNE campaigns occurred in 2009: Ghost-
Net and Night Dragon. GhostNet included "a network of over 1,295 infected hosts in 103 countries"; intruders were capable of total control of
compromised computers, including downloading data and turning on
"microphones and web cameras" (Information Warfare Monitor 2009).
Among the computer networks compromised were those of the Association of Southeast Asian Nations (ASEAN) Secretariat; the South Asian
Association for Regional Cooperation (SAARC); the Asian Development
Bank; the Dalai Lama's exile centers in India, London, and New York

City; embassies; international organizations; ministries of foreign affairs; news media; NGOs; and an unclassified computer located at NATO headquarters. A separate campaign in 2009, code-named Night Dragon, was uncovered in 2011. It targeted the energy sector which is of strategic importance to China's continued economic development. Specifically, it focused on "global oil, energy, and petrochemical companies . . . harvesting sensitive competitive proprietary operations and project-financing information with regard to oil and gas field bids and operations" (McAfee 2011; USCC Annual Report 2011). Prior to McAfee's report on Night Dragon a number of incidents targeting oil and gas were reported in 2009. If these incidents are a part of the same campaign, then Night Dragon also targeted the telecommunication and financial services sectors and attempted to map the U.S. energy grid, possibly leaving behind programs that would enable the intruder to disable services on demand (USCC Annual Report 2009; Gorman 2009). China's capability in this regard is enhanced by the universal designs of SCADA and ICS systems. The systems used by U.S. critical infrastructure are available worldwide, including in China, and utilize the same architecture, default vendor passwords, source code, and training (Weiss 2010, 57, 212).

There were numerous isolated incidents of Chinese CNE targeting multiple countries in 2009. These include the Australian Prime Minister's e-mail and mobile phones, the German Foreign Office, laptops stolen from German business travellers visiting China, the Philippines' Department of Foreign Affairs, the Intranet of South Korea's Finance Ministry, and the office of a U.S. senator (Krekel 2009; ONCIX 2011; USCC Annual Report 2009). The trend of combining insiders and CNE, noted in 2005 and 2006 incidents, continued in 2009. Three Chinese citizens employed in foreign companies used their positions to electronically steal sensitive data and attempted to use it to obtain jobs with competitors in China. These include product data downloaded to 170 CDs from a German company, "information on organic light-emitting diodes" transferred via e-mail and USB flash drive from the DuPont corporation, and "160 secret formulas for paints and coatings" valued at $20 million transferred to a removable storage device from the Valspar Corporation (ONCIX 2011; Riley and Dune 2012).

Further complicating the picture of what constitutes development of China's CNE, a series of cyber attacks occurred around the Fourth of July in 2009. Having coincided with the U.S. Independence Day, there was speculation that the attacks were symbolic. DDoS attacks targeted 14 U.S. government websites, including the Department of Homeland Security, Department of Transportation, Federal Trade Commission, Secret Service, and Treasury Department. Multiple South Korean websites were also targeted, including the Defense Ministry, National Assembly, and Presidential website (Fox 2009). These attacks were attributed to North Korea, yet they remain relevant to the development of Chinese CNE in the same way that hacker groups in China remain relevant to Chinese CNE. China's involvement is not fully understood, and it adds another layer of anonymity, if China wished to use it as a proxy. According to Clarke and Knake, North Korea operates four CNO units with "600 to 1,000 KPA cyber warfare agents acting in cells in the PRC." Unit 110 is suspected of carrying out the Fourth of July cyber attacks, and it has at times allegedly operated out of four rented floors in the Shanghai Hotel in Dandong, China, along North Korea's border. Reasons for operating out of China include increased Internet access, increased access to advanced technology, and possibly increased access to Chinese CNO expertise (Clarke and Knake 2010, 27–28).

INCREASING BACKLASH, 2010–2014

The Shadow network and Ke3chang were two APT that began in 2010. The Information Warfare Monitor revealed the Shadow network in its 2010 report, *Shadows in the Cloud: Investigating Cyber Espionage 2.0.* This report represents a continuation of the Monitor's research which revealed the GhostNet APT in 2009; however, it is also a stand-alone departure. This is in part due to many of the IP addresses and methods of exploitation used by GhostNet being shut down or blocked following the publication of their 2009 report, *Tracking GhostNet: Investigating a Cyber Espionage Network.* Additionally, the Information Warfare Monitor changed its method of analysis and this time was able to analyse the exfiltrated data to make a stronger connection to Chinese state-sponsorship. For these reasons it is unclear if its *Shadows in the Cloud* report is exposing new aspects of the GhostNet campaign or an entirely new campaign which it has named the Shadow network. The Shadow network compromised multiple computers within the entire spectrum of government networks, those of the business community, and academia within 31 countries, including India's and Pakistan's embassies in Washington, the Office of the Dalai Lama, and the United Nations (Information Warfare Monitor 2010). According to the 2010 USCC Annual Report, the perpetrators used "popular free web services—such as Twitter, Google Groups, Blogspot, Baidu Blogs, blog.com, and Yahoo! Email accounts—as part of the command-and-control infrastructure for their exploits." These websites are readily available, provide redundancy, and are less likely to appear malicious to "firewalls [or] network administrators" (USCC Annual Report 2010). In the same year, U.S. IT security company FireEye released information about a separate alleged Chinese APT called Ke3chang. Its targets represent a broad set of sectors ranging from aerospace and energy, to government and consulting services (Villeneuve, Bennett, Moran, Haq, Scott, and Geers 2014). Topics used in Ke3chang's spear phishing efforts were no less diverse, having included the London Olympics and the French first lady. The campaign gained increased attention in 2013 after targeting European ministries with malicious e-mails claiming to contain information that the United States was about to intervene in Syria (Wakefield 2013).

U.S. allegations of isolated Chinese CNE continued in 2010. Two of these incidents allegedly show China's ability to manipulate the flow of traffic on the Internet. In one case China's Golden Shield censorship was extended to Internet users in the United States for a period of several days, and in another case a large portion of international Internet traffic was unnecessarily routed through Chinese gateways which could presumably enhance data theft or blocking (USCC Annual Report 2010). In further NASA developments, the U.S. Office of Inspector General investigated several data breaches at the National Aeronautics and Space Administration, leading to the "detention of a Chinese national" who compromised, according to the investigation's findings, "seven NASA systems, many containing export-restricted technical data" (Martin 2012). Furthermore, a wide range of U.S. companies reported in 2010 that "client lists, merger and acquisition data, company information on pricing, and financial data were being extracted from company networks" (ONCIX 2011).

In another legal development, Solid Oak Software Inc. filed a $2.2 billion lawsuit against the Chinese government, two Chinese software companies, and a number of well-known computer manufacturers after discovering pirated contents from its Cybersitter software. Cybersitter was designed to allow parents, schools, and businesses to block access to specific websites, filter items, and record activity (Dilanian 2012; Branigan 2010b). The stolen source code, blacklists, and even a Cybersitter Bulletin were found within Chinese software, including Green Dam Youth Escort, that was sold to 57 million Chinese customers. Green Dam Youth Export was also found to be using components of OpenCV, a library of programming functions, in violation of its license agreement (Wolchok, Yao, and Halderman 2009). Further, the Green Dam software was discovered to contain security vulnerabilities that could allow an intruder to take control of the computer, such as uploading malicious payloads through the update feature or by exploiting how it interacts with visited web pages. This is particularly relevant given that the Green Dam software was initially mandated by the Ministry of Industry and Information Technology (MIIT) to be a required pre-installation on, or included with, all personal computers made in China or sold to China. Additionally, in 2010, two individuals were charged with the theft of $40 million worth of hybrid technology from the automobile manufacturer General Motors. The couple were accused of using a combination of traditional and electronic espionage, and attempting to sell the stolen data to the Chinese state-owned Chery Automobile Co, Ltd. (Heickerö 2012, 80).

Other alleged Chinese CNE incidents in 2010 include infiltrating the e-mail account of an Associated Press employee, targeting a Taiwanese prosecutor who was handling a number of high-profile cases involving China, and the exploitation of 150 computers at the French Finance Ministry targeting information relating to the G-20 (Baldor 2013; China says hit by 500,000 cyberattacks in 2010 2011; ONCIX 2011). Further, Canada's Defence Research and Development Canada (DRDC), Finance Department, Treasury Board, and Bay Street law firms were targeted. The Bay Street law firms were involved in an attempted hostile takeover bid of the Potash Corporation of Saskatchewan by BHP Billiton estimated at $38 billion (Nortel hit by suspected Chinese 2012). Lastly, the United Kingdom's Centre for the Protection of National Infrastructure issued a warning to UK companies about Chinese espionage methods, "including giving gifts of cameras and memory sticks equipped with cyber implants at trade fairs and exhibitions" (ONCIX 2011).

Familiar targets of Chinese CNE spanning business, government, and military were again infiltrated in 2011. Google revealed that it had uncovered a widespread spear phishing operation originating from China that was targeting activists, journalists, government officials, and military personnel (China calls US culprit in global "Internet war" 2011; China says hit 2011). Once an e-mail account was compromised, that account's contact list was targeted in additional spear phishing (USCC Annual Report 2011). Intrusions into the computers of the U.S. Chamber of Commerce revealed in the following year may be linked to this operation (USCC Annual Report 2012). NASA reported that it suffered "47 APT attacks, 13 of which successfully compromised Agency computers," including total control of key computers at the Jet Propulsion Laboratory (Martin 2012). Among the data stolen were the user credentials of 150 employees which could be used to further penetrate these networks (USCC Annual Report 2012). Additionally, U.S. Internet traffic was rerouted through Chinese gateways, similar to the Internet traffic manipulation examined in 2010, and the Australian Prime Minister was again targeted, this time along with the Australian Minister for Foreign Affairs and Minister for Defence (USCC Annual Report 2011; China says hit 2011). Lockheed Martin, the primary developer of the F-35, had some of its networks penetrated, and it is believed that the intruders were able to gain access by using data stolen from RSA Security, a division of the EMC Corporation. RSA Security developed the SecurID authentication tokens used by Lockheed Martin and others employed in the U.S. defense industry for remote access to some of their networks (China under suspicion in U.S. for Lockheed hacking 2011; Hosenball and Eckert 2011; Segal 2012; USCC Annual Report 2011). Additionally, U.S. and Canadian natural gas producers were again targeted in 2010, and the Diablo Canyon nuclear facility was compromised. The group believed to be responsible for these intrusions into the energy sector also targeted law firms involved with "anti-dumping and unfair trade cases against China," including the producers of solar cells (Riley and Dune 2012).

Continuing in 2011, China was said to have preyed upon some 760 organizations by exploiting the computer networks of iBAHN. iBHAN provided digital information and entertainment in the hospitality industry (USCC Annual Report 2012). Infiltrating these systems could enable eavesdropping on Internet communications and activity, provide avenues for compromising personal computers, and to some extent track the whereabouts of hotel guests comparable to Britain's Royal Concierge program[5] (Poitras, Rosenbach, and Stark 2013). Statistics compiled by Cloud-Fare, a website performance and security company, revealed a 56 percent reduction in malicious activity on the Internet during the October First National Day, a public holiday in the PRC, suggesting that state-sponsored CNE employees had the day off from their work (USCC Annual Report 2012). An unnamed CEO in the Australian mining industry expressed concerns over Chinese CNE, and an unnamed British businessman warned universities that Chinese students studying in the UK were stealing science and technology research and implanting malicious software into university computer networks so they can maintain access following their departure (ONCIX 2011). Chinese agents posing online as a U.S. Navy admiral deceived senior military and government officials in the UK to add them as Facebook friends; this increases the probability of them opening a malicious link or file, or inadvertently passing information which could be useful in spear phishing (Protalinski 2012). Additional computer networks alleged to be compromised by China in 2011 include the International Monetary Fund (IMF); Taiwan's Democratic Progressive Party; and French manufacturer Turbomeca, which develops engines for aircraft, helicopters, and missiles (Government "may have hacked IMF" 2011; China says hit 2011; Committee on Foreign Affairs 2011).

Three new Chinese-linked APTs occurred in 2011—Nitro, Luckycat, and Icefog. Nitro's targets included 48 chemical and advanced materials companies, whose "design documents, formulas, and manufacturing processes" represented the type of intellectual property being sought (Chien and O'Gorman 2011; Arthur 2011). In possible connection with this campaign, Chinese CNE in 2011 targeted Canadian legal firms that had intelligence on prospective agreements in the chemical industry (USCC Annual Report 2012). While Nitro is noted primarily for targeting the chemical sector in 2011, it did diversify its activities in that year to cover both human rights organizations and the automotive industry (Chien and O'Gorman 2011). The Luckycat campaign was primarily focused on India, Japan, and Tibetan activists. Luckycat compromised at least 233 computers and sought information on aerospace, energy, engineering, military research, shipping, and Tibetan activists (Luckycat Redux 2012; USCC Annual Report 2012). Icefog primarily targeted high-technology, government, media organizations, the military, shipping, and telecommunications in Japan and South Korea (Kaspersky Lab 2013b).

In 2012, there was continued concern over the possibility of China embedding backdoors, kill switches, and eavesdropping capability into critical infrastructure. Researchers discovered security flaws that allowed remote control of select ZTE mobile phones and Huawei routers (USCC Annual Report 2012). These concerns, and possible links to the Chinese government, were further highlighted in a detailed report issued by the U.S. House Permanent Select Committee on Intelligence (Rogers 2012). Further, acting on advice from the Australian Security Intelligence Organisation (ASIO) the Australia government barred Huawei from bidding on construction contracts for its National Broadband Network (McDonald and McGuirk 2012). Beyond these two Chinese telecommunication companies, Microsoft reported that one in five computers it bought from China carried malware (USCC Annual Report 2012). The discovery of counterfeit parts being used in military equipment also raised concerns over embedded backdoors, because it compromises the integrity of supply chains. The U.S. Senate Armed Services Committee found counterfeit parts in multiple weapons systems, including the RQ-4 Global Hawk UAV (Levin and McCain 2012). A separate investigation in 2012 by the US Government Accountability Office attempted to purchase parts for the F-15 fighter, the MV-22 Osprey assault support aircraft, and two nuclear submarines. An independent testing laboratory found that the parts that came from China, 16 in all, were all determined counterfeit (Reed 2012b).

Multiple countries accused China of using CNE against a wide range of targets in 2012. India's air-gapped Eastern Naval Command was compromised through the use of removable media (like flash drives), the Philippines suffered a string of incidents following confrontation with China in the South China Sea, and Taiwan's National Security Bureau reported that "27,000 discrete pieces of information" were stolen through Chinese CNE over the previous seven years (USCC Annual Report 2012; Passeri 2012). Jonathan Evans, the director general of MI5, and Major General Jonathan Shaw, the head of cyber security at the UK Ministry of Defence, separately issued public statements to the government and businesses that warned of Chinese CNE (Esposito 2012; Jowitt 2013). Japan's largest defense contractor, Mitsubishi Heavy Industries, was also compromised, and Japanese investigators linked CNE targeting Japanese legislature to the PLA (USCC Annual Report 2012). Also in 2012, the security software company Trend Micro revealed the Lurid Downloader APT, also known as Enfal. The malware and C2 infrastructure it used has some loose connections to Ghostnet, the Shadow network, and Byzantine; however, it remains a distinct APT. Lurid Downloader targeted 1,465 computers in 61 countries (Villeneuve and Sancho 2011). This comprised at least 47 victim organizations, primarily targeting the embassies, ministries, and space industry within the Commonwealth of Independent States (CIS) of the former Soviet Union. Separately in 2012, The Verizon RISK Team analyzed 621 instances of data theft from multiple, primarily Western, countries. Of these, 19 percent were determined to be state sponsored, and of those 96 percent were attributed to China (USCC Annual Report 2013).

China was further accused of exploiting the computer networks of the European Aeronautic Defence and Space Company (EADS) and the German corporation ThyssenKrupp (EADS, ThyssenKrupp attacked by Chinese hackers: Report 2013; Jowitt 2013). The Airbus Group, formerly known as EADS, is an aerospace and defense corporation that produces a wide range of aircraft. The Airbus Group was also a partner in the Eurofighter Typhoon consortium, and it has connections to the development of drones, missiles, and satellites. ThyssenKrupp is one of the world's largest steel producers and develops components for the automotive industry and industrial services. In the same year German pharmaceutical company Bayer was also compromised. The Federal Office for the Protection of the Constitution (BfV), Germany's domestic intelligence agency, reported "close to 1,100 digital attacks on the German government by foreign intelligence agencies" with the majority being traced to China (Digital Spying Burdens German-Chinese Relations 2013). Presumably the BfV's use of the word "attacks" includes CNE. Of particular significance, this figure only represents 2012, and it does not include intrusions into non-government networks.

The Defense Science Board (DSB), a committee which advises the U.S. Department of Defense, issued a classified report on Chinese CNE in 2012, the details of which were leaked the following year. According to this report 24 U.S. weapons systems were compromised, including the Aegis Ballistic Missile Defense System, Black Hawk helicopter, F/A-18 Hornet, Global Hawk UAV, Littoral Combat Ship, patriot missile defense system, F-35, Terminal High Altitude Area Defense system, and V-22 Osprey (Hoffman 2013; Martinez, Meek, Ross, and Ferran 2013; USCC Annual Report 2013). Additional U.S. victims of alleged Chinese CNE in 2012 include the Department of Defense's Common Access Cards, the nuclear research sector, ships deployed at sea, and the White House Military Office (USCC Annual Report 2012; White House says cyberattack thwarted 2012). Moreover, public servants in foreign affairs and academic staff at universities were targeted in a watering hole attack which used the Council on Foreign Relations website; and a metallurgical company lost one billion dollars' worth of technology that represented 20 years of dedicated work (Weitzenkorn 2013; USCC Annual Report 2012). Media organizations proved prime targets in late 2012. The *New York Times*, the *Wall Street Journal*, and the *Washington Post* were targeted by Chinese CNE following their investigative reporting on the wealth of Premier

Wen Jiabao's family (USCC Annual Report 2013). Similarly, Bloomberg News was targeted after reporting on the wealth of President Xi Jinping's family. These incidents represent a combination of CNE and Information Operations (discussed in chapter 3), as the motive appears to be heavily influenced by a desire to control China's image (Perlroth 2013). Further, Boxun.com was hit with DoS attacks following coverage of the Bo Xilai scandal (USCC Annual Report 2012). Boxun.com is a Chinese news website, partially owned and operated in the United States, which reports on issues that are politically sensitive in China. CNE of media organizations continued into early 2013 with Facebook and Twitter also becoming targets (Arthur 2014; Reid 2013).

The year 2013 marked a shift in tone for the U.S. Department of Defense's Annual Report to Congress: Military and Security Developments Involving the People's Republic of China. For the first time the report stated that intrusions "appear to be attributable directly to the Chinese government and military," rather than implying such with caveats on the difficulty of attribution (U.S. Department of Defense 2013). High-ranking U.S. officials, including Secretary of State Hilary Clinton, Defense Secretary Leon Panetta, and President Barack Obama, raised their concerns over Chinese CNE publicly and directly with their Chinese counterparts (Baldor 2013; Pace 2013). The summit between President Xi Jinping and Barrack Obama held at the Sunnylands Retreat in Rancho Mirage California was noted for its proximity to Silicon Valley. This, along with President Xi Jinping's prior meeting with Microsoft founder Bill Gates, appeared to signify an information technology theme to the summit. NASA also increased pressure on China relating to CNE concerns by revoking the remote-access privileges of Chinese contractors and introducing a "moratorium on granting any new access to NASA facilities for individuals from China" (Klotz 2013). These actions were preceded by the arrest of Chinese national Bo Jiang, a former NASA contractor, who was accused of attempting to return to China with stolen NASA information technology. Later in the same year, the United States passed a law which prohibits NASA, the Department of Commerce, the Department of Justice, and the National Science Foundation from acquiring any information technology system "produced, manufactured or assembled" by a PRC "owned, operated or subsidized" entity—unless there is prior approval and analysis by the FBI (Flaherty 2013). Further, the U.S. Department of Justice charged three individuals from China's Sinovel Wind Group, a wind turbine manufacturer, with the electronic theft of software source code and reverse engineering of technology belonging to American Superconductor (AMSC) (USCC Annual Report 2013). Huawei's role in Chinese CNE, and the use of traditional espionage to supplement CNE, came under renewed scrutiny following the mysterious death of U.S. citizen Shane Todd (Bonner and Spolar 2013). Dr Todd was conducting semiconductor research in Singapore for the Institute for Micro Electronics in collaboration with Huawei Technologies.

Another significant increase in pressure on China over alleged CNE was the release of Mandiant's report on APT1 which directly tied PLA Unit 61398 to the CNE activity. While the report was released in 2013, the events it details began in 2006, so it is discussed in more detail above. Coinciding with its release and the international attention it received, a new alleged Chinese spear phishing campaign targeted Japanese journalists. The e-mails used a malicious attachment which purported to be the Mandiant report (Raff 2013). Japanese organizations were further targeted in 2013 by a campaign known as Operation DeputyDog, which utilized zero-day exploits and malware payloads, and was later linked to a watering hole exploit named Ephemeral Hydra (Moran and Villeneuve 2013; Moran, Vashisht, Scott, and Haq 2013). Both of these incidents have been linked to the Beijing Group that was responsible for a number of APT, including Operation Aurora and Axiom. Taiwan's Ministry of Defense also released a report in 2013 condemning Chinese CNE, claiming that China has been conducting "CNE against Taiwan for more than a decade" and possesses "a cyber army of more than 100,000 people" (Hsiao 2013). Additionally, Australia accused China of stealing the classified blueprints of ASIO's new headquarters, and using CNE to target the Australian Bureau of Statistics, Department of Defence, Department of Foreign Affairs and Trade, Department of Prime Minister and Cabinet, and Reserve Bank of Australia (China blamed after ASIO blueprints stolen in major cyber attack on Canberra HQ 2013). Lastly, an attempted BIOS attack, and penetration into a Trend Micro honeypot designed to appear as a U.S. water plant, further illustrate the capability to transition between CNE and CNA (USCC Annual Report 2013; Ingersoll 2013).

The increased rhetoric and pressure placed on Chinese CNE continued in 2014. The U.S. Department of Defense's 2014 Annual Report to Congress: Military and Security Developments Involving the People's Republic of China again stated that some CNE appears "to be attributable directly to the Chinese government and military" (US Department of Defense 2014). Moreover, the United States indicted five Chinese military members in connection with APT1 CNE "directed at six American victims in the U.S. nuclear power, metals and solar products industries" (The United States Department of Justice 2014). Later in the same year a sixth Chinese national was indicted, and in this case was taken into custody, for a CNE scheme targeting the U.S. C-17, F-22, and F-35 aircraft (Su Bin Complaint of Hacking for China 2014; Su Bin Indictment for Hacking for China 2014; U.S. Attorney's Office 2014). The year 2014 also marked the release of Novetta's report on the Axiom APT. This report continued a trend of not only directly accusing China, but also attributing groups to specific PLA Units when possible and providing extensive attribution evidence. Further, it represents new levels of collaboration between multiple IT companies and the U.S. government to simultaneously shame the CPC publicly and provide mitigation techniques for private companies (FBI warns industry of Chinese cyber campaign 2014; Novetta 2014). Taiwan officials also continued their assertion that they were the target of frequent Chinese CNE, including the theft of "information for use in negotiations with Taiwan" (China launching "severe" cyber attacks on Taiwan: Minister 2014). Other targets of Chinese CNE in 2014 include the personnel files of U.S. government employees in an attempt to identify those who had applied for top security clearance, and identification data for millions of patients belonging to Community Health Systems (New Claims of Hacking as US, China Discuss Cybersecurity 2014; US hospital firm: Chinese hackers stole patient data 2014). In the latter case no medical or credit card information was stolen; however, the data obtained could be used for spear phishing and the operation could have been a proof of concept or mapping of the network. Additionally, China was accused of intrusions into the U.S. Postal Service and the National Oceanic and Atmospheric Administration, including the National Weather Service, compromising the personal information of more than 800,000 employees and disrupting satellite data, respectively (Flaherty, Samenow, and Rein 2014).

Allegations of Chinese state-sponsored CNE persisted in 2015 and 2016 (U.S. Department of Defense 2016). In 2015, an intrusion into the U.S. Office of Personnel Management resulted in the theft of personal information on 22 million government applicants, employees, and contractors, as well as "millions of sensitive and classified documents" (USCC Annual Report 2015). Among the personal information stolen were financial histories and over one million fingerprints, which could be used to expose foreign operatives or circumvent biometrics authentication (Congressional Research Service Report 2015). This incident has been linked to two other data breaches in 2015, one involving United Airlines, and another involving medical data of 80 million customers belonging to the health insurance company Anthem of Blue Cross and Blue Shield. United is the second largest airline in the world, and it is commonly an approved airline for government employees traveling on business, including those employed within "17 different intelligence agencies" (USCC Annual Report 2015; Riley and Robertson 2015). As with previous years examined in this chapter, the full extent of Chinese CNE incidents in 2015 and 2016 will not likely be known or revealed until several years later. For example, it was revealed in 2015 that the engineering school of Pennsylvania State University had been compromised, yet this incident occurred from 2012 to 2014 (Riley 2015; Schwartz 2015). The university develops technology for the U.S. military, specializing in aerospace engineering, and the breach included data pertaining to the university's 500 academic, commercial, and government research partners (USCC Annual Report 2015). Similarly, allegations that China compromised the security of U.S. banking regulator the Federal Deposit Insurance Corporation from 2010 to 2013 was not revealed until 2016 (Interim Staff Report 2016; Lange and Volz 2016; Pagliery 2016). For these reasons, it is difficult to ascertain trends from 2015 to 2016 as the scope of potential intrusions is not yet available in open source material.

CONCLUSION

The development of China's computer network exploitation (CNE) illustrates several key points and trends relevant to China's cyber warfare doctrine. Between 1998 and 2002, newly formed hacker groups were involved in large-scale cyber riots with Indonesia, Japan, Taiwan, and the United States. Conforming to a general nationalist sentiment among the approximately 250 Chinese hacker groups, these cyber conflicts coincided with international incidents and perceived slights to China. The hacker groups' activity during these conflicts, and in subsequent years, was primarily computer network attack (CNA); however, it set the stage for Chinese CNE which became dominant around 2002. This chapter has explored over 100 isolated CNE incidents and 19 APT assigned to China through a collection of six attribution elements. China appears to have recognized the benefits of online exploitation and embraced it through the formation of approximately 20 state-sponsored CNE groups. Among its benefits, it provides a relatively inexpensive means to leapfrog decades of research and development in all sectors, and thus far, it has incurred minimal punishment. Through the Internet, China has the ability to conduct large-scale espionage operations against thousands of targets, stealthily, and on a global scale. This dispersed trickle of intellectual property may, over a period of decades, result in a "death by a thousand cuts" for Western states.

Alleged Chinse CNE spanned all sectors of business, government, and military, within 71 countries, targeting any information which would be of value to the Chinese government. Some single APTs, and even some individual isolated incidents, targeted hundreds of organizations, infected thousands of computers, and cost billions of dollars in losses. The types of damage caused by alleged Chinese CNE include advantageous competitor insight, altering data, delays in production, forced replacement of computers, increased cost, increased work restrictions, loss of data, reduced efficiency, service outages, the spread of misinformation, theft, and (unconfirmed) electricity blackouts. The PRC has also been accused of mapping computer networks and leaving behind tools to aid Chinese CNA capability. Typical methods for initial access to computer networks included socially engineered spear phishing and watering hole attacks. These, along with further exploitation after the initial breach, utilized the full range of common malware. Some of the more novel vectors of exploitation allegedly used by China during the period of 2000–2014 were imposter Facebook accounts; manipulation of large flows of Internet traffic; mobile devices, including phones, cameras, and flash drives; the ability to turn on a target's webcam and microphone; eavesdropping on online conference calls; targeting the providers of secure identification and encryption in order to breach the systems of their clients; and a variety of embedded backdoors.

There are a number of recurring themes in alleged Chinese CNE. The Huawei Technologies Company was accused by multiple countries of facilitating the use of embedded backdoors by the Chinese government in 2003, 2005, 2007, 2008, 2012, and 2013. Similarly, Chinese computer manufacturer Lenovo was accused of implanting exploits in its merchandise in 2006, 2007, and 2008. Two separate studies in 2012, by the U.S. Government Accountability Office and the Senate Armed Services Committee, raised concern that counterfeit parts and compromised supply chains may be furthering China's ability to embed backdoors. Additional incidents in 2008, 2010, and 2012 demonstrated pre-installed malware within Chinese computer products. Another trend within Chinese CNE is the use of insiders and traditional espionage to complement cyber espionage (HBGary 2011). This was demonstrated by three European cases in 2005; a Ford Motor Company incident in 2006; the physical theft of laptops in 2008 and 2009; three cases involving LED, chemical, and paint companies in 2009; Su Bin's alleged attempts to gather information on military aircraft between 2009 and 2013; a General Motors Company incident in 2010; and the disputed death of Shane Todd in 2013.[6] Other trends in alleged Chinese CNE include recurring targets. Intrusions into NASA's computer networks have been continually occurring since 2002, including some incidents which belong to the Titan Rain and Avocado APTs. Additionally, information pertaining to the F-35 fighter aircraft has been under a sustained, yet seemingly scattered, CNE effort since 2007.

International condemnation of alleged Chinese CNE, and pressure placed on China to cease CNE activities, has taken place in multiple forms and is on the rise. Public accusations by government officials had a sharp increase around 2007 and continued through 2014. These were individual accusations from multiple countries, rather than a single APT report which listed multiple countries as victims. Around 2010, accusations of CNE began to convey greater confidence in attributing intrusions directly to the Chinese government. Between 2010–2014, government offices and IT security firms began releasing a higher quantity of statistics, including the annual number of intrusions, estimated value and percent of losses, attributable to China. There also appeared to be a new trend, particularly within the United Kingdom and the United States, of issuing warnings directly to business leaders of ongoing Chinese CNE activity, possibly representing an attempt to improve intelligence sharing. A number of high-profile reports on alleged Chinese APTs were made public during this particular period, including Google's revelation of Operation Aurora and Mandiant's report on APT1. These represent new levels of boldness in assigning attribution, collaboration, and transparency, which might encourage other companies to come forward with information about computer network breaches.

Further, 2010–2014 revealed an increase in attempts to use legal action against Chinese CNE through arrests, bans, and lawsuits. In connection with intrusions into the computer networks of NASA, a Chinese national was detained in 2010 and another was arrested in 2013. In 2014, five Chinese military members were indicted in relation to APT1 and a sixth, separate, Chinese national was taken into custody for a CNE scheme targeting U.S. aircraft. While these may not result in convictions, they represent a hardening stance and vigilance on the part of the United States. In 2012, Australia prohibited China from taking part in the development of its National Broadband Network. This coincided with, and drew further attention to, the investigation of the U.S. Permanent Select Committee on Intelligence into Chinese telecommunication companies Huawei and ZTE. In 2013, NASA barred remote access to its computer network by Chinese contractors, and the United States passed a law prohibiting the purchase and use of unauthorized Chinese-made information technology within select U.S. organizations. Lastly, lawsuits in relation to Chinese CNE were filed in 2010 and 2013 on behalf of Solid Oak Software Inc. and American Superconductor, respectively. A rise in accusations occurring roughly between 2010 and 2014, along with increased media exposure, could be the result of enhanced awareness of Chinese CNE and a decision to take a stronger stance toward China. It could also reflect a more profound appreciation of China's overall rise in power and the emergence of the Internet as a vital feature of contemporary society. Despite increasing publicity, the benefits China has allegedly gained by conducting CNE appear to far outweigh the costs. For this reason, it is probable that Chinese CNE activity will continue to increase rather than decrease.

NOTES

1. APT and malware are commonly named after recurring words or phrases found within the code or infrastructure of the aggressor.

2. For all incidents discussed within this chapter, China is either directly stated as being responsible, or it is strongly implied, by the source.

3. The court case concerned a former imperial Japanese soldier, Azuma Shiro, who admitted to war crimes against China. Shiro's appeal against the court's denial of the validity of his account failed according to a ruling by the Japanese Supreme Court in 2000 (Nanjing massacre diary author Azuma Shiro died 2006; Wu 2006).

4. The Nortel campaign began in 2000, but it was revealed in 2012. Similarly, the Byzantine campaign began in approximately 2002 but was not made public until much later. Byzantine also has an unclear connection to a different APT campaign beginning in 2006, hence it will be explored later in this chapter.

5. The Royal Concierge Program is an alleged computer system, operated by Britain's Government Communications Headquarters, which tracks the hotel bookings of foreign diplomats at approximately 350 high-end hotels around the world. This information, including advanced warning of a diplomat's arrival, could increase the ability of British intelligence agents to conduct further CNE, human intelligence, surveillance, and wiretaps (Paterson 2013; Poitras, Rosenbach, and Stark 2013).

6. China was accused of further traditional espionage during the period 2000–2014 (Luard 2005); the cases noted here only represent those which appeared to directly supplement CNE.

THREE

Information Operations

Information Operations (IO) entails utilizing and manipulating information online, without hacking, to influence actions.[1] This encompasses a wide range of activities including censorship, propaganda, and psychological operations (PSYOPS). It can be defensive or offensive, used during peacetime or war, and aimed at domestic or foreign audiences. Compared to other states, China has much more control over Internet access within its borders and requires less hacking, therefore IO has a large role in China's cyber warfare doctrine. Further, it is essential to examine IO, because it is intertwined with Computer Network Operations and Net-Centric Warfare. This connection will be demonstrated further in chapter 5 when Chinese cyber warfare is applied to the Taiwan issue. As with other Chinese concepts, such as unrestricted warfare and comprehensive national power, China takes a holistic approach to cyber warfare (Andress and Winterfeld 2011, 43–44; China's Military Strategy 2015; Gady 2015). *The Science of Campaigns* prescribes "a combination of the orthodox and unorthodox" activities, while *Science of Military Strategy* instructs to "analyz[e] the enemy and our situations systematically and in their entirety" (Anderson and Engstrom 2009). As such, this chapter will investigate the difficulties in mapping the structure of Chinese IO and arrive at a clearer understanding from which to work. China's attempts to influence foreigners via the Internet, and its attempts to defend against undesirable messages targeting China, will be examined in order to demonstrate what is possible with IO and the challenges China faces.

DIFFICULTIES IN CATEGORIZING IO

Identifying all elements that fall within the branch of IO and placing them into an organized structure poses several problems. Cyber warfare literature which discusses using the Internet to one's advantage without hacking includes terms such as coercion, censorship, deception, influence, media, perception, propaganda, psychological, public opinion, and shaping; often followed by the word operations or warfare. There is significant overlap between these terms. Attempting to categorize them, such as through identifying their relevant government departments, can create additional overlap. For example, ". . . in 2003 the Communist Party's Central Committee and the Central Military Commission approved a new warfare concept for the PLA, the 'three warfares' . . ." (Wortzel 2013, 151). The Three Warfares (sān zhǒng zhàn fǎ;—三种战法) are categorized as public opinion (or media), psychological, and legal (Anderson and Engstrom 2009; Deptula 2011; Krekel, Adams, and Bakos 2012; Thomas 2009, 112, 221–23; U.S. Department of Defense 2010; Wortzel 2014). One problem with these distinctions is the overlap in meaning; media can be used to conduct PSYOPS, and PSYOPS can affect public opinion. Further, while the Three Warfares are often discussed in Chinese cyber warfare literature, it is important to note that they are not cyber specific. The Three Warfares include public opinion (or media), psychological, and legal outside of the cyber realm. In some cases, all that separates one element, like propaganda, from being cyber or non-cyber is the means by which it is transmitted (electronic versus print); however, IO does contain other unique methods of influence which do not have an offline equivalent.

Additionally, the Three Warfares can be applied to all three branches of cyber warfare, not only IO. For example, CNO can be used to deface web pages, spread viruses, or alter data on a network. These can weaken morale, intimidate, or erode trust in information networks, making them CNO-enabled PSYOPS. Some examples of NCW-enabled PSYOPS include using high-tech camouflage to trick satellites, and using UAVs to broadcast a message or drop talking leaflets (Thomas 2004, 96–98). Turning to the media component of the Three Warfares, the "jamming of international broadcasts, including both Radio Free Asia [RFA] and the Voice of America [VOA]" in Tibet and Xinjiang are an example of NCW-enabled media warfare (Access to Information and Media Control in the People's Republic of China 2008). Further, allegations that China used CNO to intimidate U.S. media organizations, and to "gain warning about negative media coverage of China before it [was] published," appear to demonstrate a combination of CNO-enabled PSYOPS and CNO-enabled media warfare (USCC Annual Report 2013). Acting on this advance warning about negative media coverage by publishing pre-emptive articles online to counter the upcoming criticism would be IO-enabled PSYOPS and IO enabled media warfare. The legal, or lawfare, component of the Three Warfares is also applicable to IO; however, it is more of a supplement than a defining characteristic of this branch. As with PSYOPS and media warfare, lawfare applies to all three branches of cyber warfare.

In determining whether the application of lawfare is an IO topic it is necessary to ask whether the intended outcome is going to affect an IO issue. China's push for international legal efforts against the weaponization of space are NCW-related lawfare, whereas laws requiring foreign companies operating in China to comply with domestic censorship practices are IO-related lawfare. IO itself can be used to enhance lawfare capabilities in multiple realms by using online influence to aid in the adoption of favorable laws.

Rather than try to force IO to correspond with an existing structure, a more productive approach is to examine its constituent elements and to see if one can detect a pattern. The Internet is a communication tool, and the various non-hacking techniques—from media to psychological and censorship to propaganda—are concerned with conveying a message. The purpose of the message is to influence a target's opinions and subsequently their actions. Information Operations Exploitation (IOE) entails using online information to one's advantage, or more specifically, disseminating messages (true, false, or misleading) which can induce a preferable action. Information Operations Defense (IOD) protects against undesirable messages, by blocking them, changing them into preferred messages, or using them to undertake an opposing action such as an arrest.[2] Common examples include propaganda and censorship, respectively. This model also applies to less obvious examples at the extremities of the spectrum. Mundane activity such as a business creating its own website, the government judging a lower-ranking Party official's performance based on online opinion and information, or the cloning of popular Western websites for a Chinese audience (if done without hacking) are all related to online messages influencing actions.[3] More exotic examples include virtual dead drops, which entail two people using the same e-mail account to pass messages by saving them as drafts; or steganography, such as hiding information and communications within a publicly available image file or Microsoft Word document (Fritz 2008; Mandiant 2013; Thomas 2009, 111–12). With these the message is hidden, aimed at a select audience, yet available online with the intent to influence action. In some cases, undesirable messages can be beneficial in that they provide actionable intelligence. There can also be complexity within the messages used by IO, such as one message causing two opposite, yet desirable, actions among dual targets. A message that intimidates or demoralizes a foreign audience could also encourage a domestic audience through nationalism, creating a type of win-win message. With the Internet these messages can be in many forms, such as news articles, social media, mp3s, videos, and image files. This way of conceptualizing IO bypasses the difficulties of overlapping terms, and it can help reveal new possibilities rather than continually rediscovering or reorganizing old ones.

While the above two forms of "messages influencing actions" are useful for verifying and analyzing examples of IO, further delineation is needed for the purpose of this book. At first IOE and IOD seem comparable to CNO's CNE (or CNA, due to its offensive nature) and CND. CNO is largely concerned with attacking or exploiting other states and defending against attacks or exploitation coming from other states. Similarly, it seems IO should be largely concerned with communicating preferred messages to other states and defending against undesirable messages coming from other states. However, there are some difficulties with this, because there is much more domestic-foreign overlap in IO, due to the nature of globalization and the Internet.[4] The Central Publicity Department (CPD), formerly the Central Propaganda Department and often referred to simply as the Publicity Department, is responsible for propaganda targeting a domestic audience, while the State Council Information Office (SCIO) is responsible for propaganda targeting abroad. China traditionally refers to these types of propaganda as internal and external, respectively (China's Propaganda and Influence Operations, Its Intelligence Activities That Target the United States, and the Resulting Impacts on U.S. National Security 2009; Shambaugh 2007). However, the distinction between them has become blurred with the Internet. For example, there are a large number of foreigners within China or abroad who read Mandarin, and there is the large Chinese diaspora, making it hard to send split messages. Further, internal IOE can influence Chinese people to conduct external IOE. This is unique to IO; under CNO it is unlikely that China would want to attack its own people in order to influence them to attack a foreign state. While this is conceivable in CNO, it does not warrant more than a theoretical mention in the context of this book; whereas there is a vast amount of literature detailing internal IOE and a high frequency of inspiring people to write preferred messages online that reaches foreigners. Other ways in which internal IOE is relevant to international relations include its use to counteract external IOE conducted by foreign states, or to quell nationalism from escalating into an international incident. The latter could damage China's soft power, incite damaging retaliatory attacks, or force China into confrontation with a foreign state.

Internal and external overlap is also seen within IOD. The Golden Shield Project (Golden Shield), also known as the Great Firewall, is predominantly known for its use in internal IOD, and it is owned and operated by the Ministry of Public Security (MPS). Yet the Golden Shield plays a vital role in external IOD, and many of the processes used for external IOD are the same as those used for internal IOD, therefore there is some inevitable crossover when attempting to examine the Golden Shield for external IOD. Further, a foreign state can influence Chinese people to use IOE against their own government. Chinese external IOD attempts to stop the initial foreign message before it can have this influence, yet when looking at IO comprehensively, the next level of defense would include internal IOD. CNO does have an equivalent, such as a virus created in another state being used by a Chinese person; however, tracing the origin of code seems more tangible than tracing the origins of influence. Barriers to use, such as the technological limitations, the perceived consequences, and feelings of personal responsibility are greater with spreading a virus than spreading an opinion. Despite these limitations and overlap, the scope of this book is limited to a focus on international relations, therefore the following sections will emphasize external IOE and external IOD, and only note internal when it is intertwined with external, rather than a full exploration of all aspects of internal. Similarly, this research will not delve into the details of psychology, marketing, and advertising.

A final difficulty to be addressed is trying to identify the government, or organizational, structure in place for conducting IO (see figure 1.1). Detailed responsibilities and chain of interagency command for specific operations are considered state secrets (Shambaugh 2007). This lack of transparency is in itself a useful form of deception making it difficult for a foreign state to create an effective opposition and disguising the purpose or role of officials at various events and interactions with foreigners. Although it may also reflect an inefficient structure and competing interests within, so it is uncertain whether the benefits are worth the costs. Multiple reports place IO under the purview of SCIO with supplemental actions being taken by a variety of ministries (China's Propaganda and Influence 2009; Lawrence and Martin 2013; Rumi 2004; Shambaugh 2007). General descriptions of these departments' responsibilities posted online by the Chinese government support these claims as they list several IO-relevant areas (The Organizational Structure of the State Council 2003). However, a report prepared by the Northrop Grumman Corporation for the U.S.-China Economic and Security Review Commission (USCC) in 2012 places IO under the joint responsibility of the 3PLA and 4PLA coordinated by the Department of Informationization and the Information Assurance Base (Krekel, Adams, and Bakos 2012). One possible reason for this discrepancy is that the Internet is a communication and media device, while at the same time it holds significant military application. Traditionally the military would hold a greater role in PSYOPS while the media would deal more with censorship and public opinion; however, the two have become blurred with the Internet, just as peacetime and wartime operations have become blurred. Northrup Grumman's report to the USCC might reflect a shift in area of responsibility toward the PLA or overlapping and competing interests. Meanwhile, Larry M. Wortzel's 2013 book, *The Dragon Extends Its Reach*, places IO under the authority of the General Political Department (GPD) in close partnership with SCIO. This book was written after the Northrup Grumman report, and Wortzel is a commission member on the USCC board, therefore it is probable that this information is more up to date. It shows that the GDP works closely with SCIO, bridging the gap between the State Council and Central Military Commission (CMC) which could represent the beginning of a more efficient structure.

EXTERNAL IOE

China seeks to use IOE to aid in its continued modernization, avoid direct confrontation, maintain CPC control, reassure its neighbors and the world of its peaceful rise, and reduce Western influence. Chinese history is replete with examples of using deception and perception management to gain advantage, from Sun Zi's *Art of War* and the *Thirty-Six Stratagems* to classic literature like *Romance of the Three Kingdoms, Outlaws of the Marsh*, and *Journey to the West*. Mao Zedong sought to minimize the perceived threat posed by the CPC and successfully utilized propaganda, while Deng Xiaoping instructed to "hide our capabilities and bide our time; be good at maintaining a low profile; and never claim leadership" (Anderson and Engstrom 2009; China's Propaganda and Influence Operations 2009). China's rich history shows a predisposition to IOE, yet China does not highlight these traditions within external IOE itself, because that would lessen its effectiveness. Instead, China's external IOE focuses on other aspects of their history: a revival of Confucian ideals, Hu Jintao's Harmonious Society, and the naval voyages of Zheng He (Dotson 2011). This may be deception, or it may reflect the true beliefs of the current Party, either way it serves as useful IOE by attempting to generate favorably foreign opinions and alleviate what some perceive as the China Threat. Beginning in the 1990s China issued instructions to boost "external publicity through the Internet" and assigned public diplomacy with five main objectives:

> (1) more strongly publicizing China's assertions to the outside world; (2) forming a desirable image of the state; (3) issuing rebuttals to distorted overseas reports about China; (4) improving the international environment surrounding China; and (5) exerting influence on the policy decisions of foreign countries. (Rumi 2004)

Common examples for achieving these objectives include official statements to online media organizations and the posting of government White Papers. Evidence of China implementing IO strategies can be seen with Tsinghua University teaching government officials "how the media works" and attempting to move away from the "static phrases, slogans, and warnings" characteristic of China's past (Access to Information 2008). Additionally, military courses offered at Wuhan University include "an introduction to U.S. and Taiwanese social information systems," suggesting that China has recognized the benefits of utilizing social networking externally (China's Proliferation Practices, and the Development of Its Cyber and Space Warfare Capabilities 2008). External IOE is most effective when personnel have a thorough understanding of the language and culture of their target audience, at both the state and online community levels.

China can enhance the effectiveness of its messages by directly or indirectly enlisting the assistance of foreigners in its external IOE, in a policy called "using foreign strength to promote China" (China's Propaganda and Influence 2009). One technique is to interview, or hire as a foreign correspondent, Western celebrities from the entertainment, political, or academic spheres who align with the CPC's preferred message. These individuals add credence and visibility through their standing, and clarity through their fluency and understanding of the subtleties of language and culture. In some cases, they may not need to be hired, they are already producing material such as news articles or academic research which aligns with the CPC's preferred message. This material can be promoted online, possibly through the aid of another type of celebrity, social media "power users," or trendsetters. These individuals can popularize web links, generating traffic in the tens of thousands, which is then picked up by traditional media outlets. This activity also improves the material's rank in search engine results, thus perpetuating the spread of that viewpoint, such as the likelihood of it being referenced in subsequent academic papers. This is accomplished through a deep understanding of online culture, history, and technology, and it is possible to purchase such a service (China cyber-gangs use "vast underground network" 2014; Lawrence and Riley 2013). Another technique is to provide funding for think tanks and university scholarships which can set research agendas and topics. Further, "granting access to archives, research opportunities, interviews," visa approval, and giving guests special treatment can all result in favorable online content, "sometimes consciously, and often unconsciously" (China's Proliferation Practices 2008). This is not necessarily insidious; it is good practice. Non-celebrity foreigners are also valuable in putting out external IOE, because of their language and cultural fluency, along with appearing less controlled by virtue of being non-Chinese. Mobilizing the Chinese diaspora is another type of foreign assistance. Online support for China during the 2008 Tibetan unrest and Olympic torch relay resulted in the organization of real-world demonstrations and protests, and included individuals "attaching a red heart moniker next to the word China to their avatars" (China's Proliferation Practices 2008).

China may wish to tap into the power of a broad range of domestic Internet users, those who are not celebrities, nor skilled hackers, yet have wide-ranging knowledge of the Internet through frequent use. Government officials have hired "tens of thousands" of Internet commentators, often known as the 50 Cent Party, who attempt to influence public opinion online in favor of CPC policy[5] (Bristow 2008; Thomas 2009, 90). Comments posted online in message boards or attached to news articles, blogs, videos, and social media might reflect genuine support for a CPC-endorsed point of view, or they could contain misleading or false information[6] to try to persuade people. Democratized news, where comments can be voted up or down, or news articles, videos, and mobile apps (including malicious ones) are displayed more prominently based on votes, can be artificially inflated using scripts or multiple accounts belonging to the same user (sockpuppets) (China cyber-gangs 2014; Cuban 2008). Xinhua could attempt to raise its international popularity in the way that Russia Today and Al Jazeera have risen. They could purport to show news and opinion that is unavailable by Western media, possibly showing controversial and self-critical pieces as a way to build an initial user base. Flooding YouTube with news clips, artificially inflating the view count, and posting positive comments would increase their presence. Even if a more experienced audience can see through these tactics, it can make a false impression on casual Internet users or the young. For example, a student researching a topic for presentation may replay one of these videos in her class because it was readily available and the view count and comments seemed to demonstrate credibility, thus the message spreads. While there is value in quantity, quality still matters. For this reason, China also needs fully trained and dedicated external IOE operatives, beyond common Internet users, and a coordinated strategy to maximize effectiveness.

A more advanced and fully employed commentator can be trained in PSYOPS, mob mentality, propaganda, logical deterrents, and key topics of discussion. They could subtly, or aggressively, point out flaws in online commentary which does not support the CPC's views, and present researched and well thought-out counterarguments. Subtle techniques, such as self-deprecating humor, can sway a crowd's emotions and train of thought. This is where the previously mentioned psychology, marketing, and advertising come into play; and they might be desirable academic degrees for a career in IOE. Beyond commentating or promoting, operatives can also monitor Internet activity and set up online polls to help judge public opinion for crafting appropriate messages. There is a fine line between soft power and propaganda; if not carefully orchestrated, external IOE can do more damage than good. For example, on multiple occasions China has been accused of using Photoshop to digitally alter photos in its favor (Pasternack 2008; Yue and Yue 2008). For this reason, external IOE should place an emphasis on finding, and clandestinely promoting, favorable media from third parties when possible. Similarly, promoting messages of truth poses less risk and can still have beneficial results; although a combination of techniques is optimal. Since IO is restricted to the Internet, and is available to anyone, IO deals primarily with mass opinion. However, in some cases these messages, while visible to the general public, are intended to influence specific foreign leaders and officials. For example, statements through online media can indicate how China would respond to specific foreign actions. External IOE also seeks to transition seamlessly from peacetime to wartime. It can be used to build justification and support for war, or boost nationalism during conflict such as posting a video of precision strikes on YouTube. As discussed in chapter 2's *Rise of Hacker Groups, 1998–2002*, and seen in subsequent conflicts involving Israel and Russia, cyber warfare has become a supplement to wars and crises, and IO plays a role in this.

IOE can be used as a means to control, or as a way to prepare, an unknowing militia. In one view, patriotic Internet users "are treated as useful idiots by the Chinese regime," and China "has figured out a rough method, using the propaganda apparatus, to shape [their behavior]" (China's Proliferation Practices 2008). China can use skilled IOE operatives to influence and direct this larger crowd, using them as a political tool. For example, internal IOE can be used to organize protests and CNA to denounce Japan's lack of remorse for WWII atrocities, criticize Falun Gong followers, or rally support for the One China Policy (Faiola 2005). The 2003 Information Operations Roadmap notes that PSYOPS and manipulating the thoughts of populations through the media and Internet requires constant observation during peacetime, otherwise in the event of conflict, a state would not be sufficiently engrained into the information culture to utilize them fully. This includes knowledge of websites' histories and the rapid evolution of memetics, pop culture, slang, and subcultures. Controlling this sort of "mob" is not a skill that can be quickly obtained. Project Chanology gives insight into how these non-hacker Internet users can come together toward a common goal of disruption using the rapid growth of available Internet capabilities (Fritz 2008, 35–37). While Anonymous is a Western example, it is a notable case study and comparable to China's Honker Union or Red Hacker Alliance. In the case of Anonymous, humor is used to unite and to obfuscate logic and responsibility. Credence within the group may come from inside jokes and creativity, rather than sound information—even revelling in their own failure. While China seeks to harness the power of these emergent communities through IOE, they also pose a non-traditional security threat, which China must understand in order to construct an effective defense.

EXTERNAL IOD

The Golden Shield plays an integral role in IO defence. China's "Internet filtering system [the Golden Shield] is the most sophisticated in the world" (USCC Annual Report 2005). It uses regularly updated lists of IP addresses, URLs, DNS, VPNs, packets, and keywords which can be blocked or filtered (held for review or automatically altered) at any of China's three major international gateways located in Beijing, Shanghai, and Guangzhou. To paraphrase Richard A. Clarke, it is like trying to prevent a terrorist attack in Manhattan by inspecting traffic entering the four tunnels that connect to the island, rather than trying to defend every building (Clarke and Knake 2010, 161). However, these gateways are only one level of the Golden Shield. ISPs and individual businesses, such as hotels, Internet cafes, individual websites, and universities, employ additional blocking and filtering. In some cases, they can be held responsible for their customers' activity and are required to check identification and keep detailed records. China also relies on "many tens of thousands of censors, working inside Chinese social media firms and government at several levels" to manually check content and remove questionable material (King, Pan, and Roberts 2014). In 2009 China attempted to make the Green Dam Youth Escort content control software a mandatory pre-installation on all computers sold in China or imported from abroad. This was later made voluntary, but it shows China's interest in improving control by placing another layer of defense at the individual computer level. Individual users can also be influenced to comply with government controls either willingly or through fear of the consequences, resulting in self-censorship. There are some ways to bypass the Golden Shield, such as proxy servers or virtual private networks (VPN); however, "the vast majority of Internet users will not, or cannot, make the effort to try to get around censorship . . . these can be technically challenging, insecure, slow, and unreliable . . . many of these services are unencrypted and so easily monitored" (Access to Information 2008).

Beyond a lack of transparency on the part of the government and frequent updates to blacklists, the Golden Shield appears unpredictable and sporadic due to its multilayered approach. Certain keywords and websites may be blocked briefly or for long periods of time. The block might be lifted during visits by foreign dignitaries or after the dates of sensitive anniversaries and Party meetings have passed, and blocks may only apply to certain cities. Specific IP addresses can be assigned to have less restriction, either for those who have been granted approval, or given to foreigners without their knowledge. For foreigners this can give the impression that all computers in China have that same access, or give a false sense of security, causing them to be more open in their activities and communication without realizing they are being monitored. Another reason for discrepancies in research stating what is, or is not, censored is the wide array of censoring practices and software used at the mid-level of the Golden Shield (King, Pan, and Roberts 2014; Qiang 2005). Different businesses use different software, in addition to the statewide filtering, which will yield different results. Different locations also have an interest in blocking information that is unique to their area, such as issues concerning the local Party's image; and self-censorship will vary among individuals and regions. Tactics also vary depending on circumstances or who is in charge. In some cases, it might be deemed that allowing users to vent their frustration online can prevent escalation of an incident, where-as other times the opposite is true. An alternative tactic is to flood comments, websites, or topic areas with the opposing view, so without removing something it has been made to look like a minority opinion. A 2008 hearing before the USCC reveals a win-win type of IO:

> Access to websites such as YouTube, Blogspot, and Wikipedia have been unblocked these past months . . . Chinese authorities probably realized it made them look bad on the international scene and those websites have a limited audience in China, so blocking them was not worth the trouble. The BBC news website in English was recently unblocked, but the BBC Chinese news website was not . . . Xinhua, the state news agency, is releasing an increasing number of stories only in English about previously taboo topics, such as peasant's riots, in a clear attempt to show the rest of the world that China is opening up, while keeping this information out of reach of most Chinese citizens. (Access to Information 2008)

Censorship of only the Mandarin language is also seen in the blocking of key word searches. In 2011, microblogging websites blocked searches for the name "Hong Bopei." This was the Chinese name for the United States Ambassador to China, Jon Huntsman. While searches for his English name continued to provide search results, searches for his Mandarin name yielded the message "according to the relevant laws, regulations and policies, the search results cannot be shown" (Page 2011). This incident was allegedly in response to photos and a video which had surfaced online showing Huntsman outside a Beijing McDonald's restaurant during a pro-democracy protest which called for a Jasmine Revolution in China.

This uncertainty and variation makes it difficult for an opposing force to develop an IOE campaign against China, and it limits criticism of the Golden Shield by foreigners, because it is fluid. A report might claim that a specific website or term is blocked, but then there will be others who will dispute this because it is not the case in every city or on every day. By avoiding detailed official statements of what is not allowed, China avoids some international scrutiny and condemnation, such as when the government issued its "11 Commandments of the Internet" in 2005 (Reporters Without Borders 2005). When an official statement is made, keeping terms vague allows China greater flexibility when deciding whether to punish various online activities, and the uncertainty causes more self-censorship. The root concern for China in its IOD is maintaining CPC control. The following is a list of key topics and words frequently blocked by the Golden Shield, which helps identify China's perceived threats: *Arab Spring, collective action (protests, demonstrations, etc.), corruption, criticizing Party leaders (or publishing their family and income details), defending the Dalai Lama, divide between rich and poor, ethnic unrest, Falun Gong, forced relocation or land seizures, gambling, human rights abuses, mining incidents, pollution, poor construction practices such as revealed by collapsed schools during the Sichuan earthquake, pornography, pro-democracy, outbreaks of infectious disease (avian influenza H5N1, HIV/AIDS, SARS), Taiwan independence, Tiananmen Square incident, Tibet independence, terrorism (aid in), Xinjiang (East Turkestan) independence* (Access to Information 2008; King, Pan, and Roberts 2014; Qiang 2005).

Annual surveys conducted by the Chinese Academy of Social Sciences "suggest that Chinese attitudes support some varieties of government censorship" (Access to Information 2008). This might reflect a more communal culture and fear of oppression by foreign powers following the narrative of the Century of Humiliation (from the First Opium War in 1839 to Communist victory in 1949). While not as extensive, Western countries do regularly block Internet content dealing with sexual exploitation of children, various forms of hate speech, and some content that aids terrorism and criminal activity. Further, the U.S. PRISM program reveals a desire by some in the United States for increased surveillance. In terms of IOD effectiveness, the Golden Shield offers China many advantages which are currently unavailable to the West. At the same time China must carefully observe foreign press and consider possible outcomes, rather than blanket blocking and confrontational denouncements. Here the line between IOE and IOD begins to blur comparable to active defense (see the introduction). China must listen to and understand the foreign audience in order to construct effective counter messages. In some cases, which are deemed highly sensitive, Xinhua is the only media outlet allowed to report on the matter, so that it can be closely controlled; all other media outlets are confined to reprinting the Xinhua article (China's Propaganda and Influence 2009). In some cases, minimal responses to foreign criticism, or no response at all (not taking the bait) is preferable. Western media is often seen in China as being anti-Chinese, in part due to the high percentage of foreign articles which focus on the negative or threatening aspects of China, as opposed to positive stories covering the country's rich history and alleviation of poverty. While nationalism might help distract from censorship concerns domestically, it is also an IOD concern for the CPC because it could escalate beyond its control. For example, during the 2008 Tibetan unrest there was widespread anger toward foreign media:

> Several foreign news media—especially those with websites enabling visitors to post comments—were flooded with messages repeating the government propaganda word-for-word. Many foreign correspondents of such Western media such as CNN, BBC, and *USA Today* received death threats after their personal information was posted online. The website antiCNN.com was also launched. (Access to Information 2008)

In addition to states, China must defend against foreign messages from non-state actors, particularly terrorist groups. In 2013, China supported the UN Security Council's anti-Internet terrorism resolution 2129 (Prevent cyberterrorism 2013). China labels terrorism as one of the "three evils," along with separatism and religious extremism. Recent attacks have shown terrorists striking beyond Xinjiang and Tibet, with the 2013 Tiananmen Square attack and the 2014 Kunming railway station massacre. Information gained on the Internet can yield maps of installations, bus schedules to and from those installations, operating hours, photographs, telephone/e-mail directories, and so forth. Much of this may be considered non-sensitive information on its own, but when pieced together it can reveal a picture which may have been deemed classified. In this regard, China has pressured the United States to reduce the image quality of satellite photos provided by Google Earth, particularly those showing Chinese government facilities, stating that the United States "could be held responsible if terrorists used that information" to conduct an attack (Glanz and Markoff 2010). Additionally, the Internet's ability to identify specific groups based on ethnicity, belief, or affiliation has enhanced the ability of terrorists to recruit and target. This can be used to identify individuals who may possess specialized access or knowledge, who can be targeted for recruitment or an e-mail phishing campaign. In terms of recruitment, many terrorist organizations operate their own websites, complete with propaganda, donation collection, and information on how to join their cause. China closely observed the U.S. operations in Iraq, where Sunni insurgents have used the Internet to post articles and video which undermined coalition forces by glorifying terrorism, demonizing the coalition, and promoting their interpretation of events (Carfano 2008). The Chinese Foreign Ministry has stated that the East Turkistan Islamic Movement (ETIM) has "published many audio and video products to stir up so-called Jihad against the Chinese government in recent years" (China hails UN's anti-Internet terrorism resolution 2013). It has been reported that Islamic militants from Xinjiang have joined the Islamic State in order to gain training which they can use to conduct attacks in China (Hui and Wee 2014). Due to the global nature of the Internet, authorities have difficulty in shutting down these sites as the web host may be located in foreign states with varying laws, and alternative hosts can be set up relatively easily if one is shut down. Rather than try to shut down the source, the Golden Shield attempts to prevent these sites from being accessible in China.

China faces several difficulties in maintaining IOD through the Golden Shield. Readily available free web sources, such as blogs, photo uploading, video uploading, podcasts, torrents, and social networking sites have given powers to individuals that were once restricted to large media outlets. Attempts to block these new forms of distribution were seen during the 2009 Ürümqi riots, the 2008 Tibetan unrest, the 2003 SARS outbreak, and numerous local protests. While this is a primary concern for internal IOD, it also relates to external IOD, because average citizens from other states can take part. Low cost, low entry level, the removal of distance, and a high level of anonymity reduce barriers for someone deciding to participate. Redundant communication platforms and the rapid pace of technological change also make it difficult to defend against. Foreign states acting within China have been using IOE against China, such as the U.S. embassy in Beijing using Twitter to report on air pollution, often contradicting the Chinese government's reporting (Wong 2013). Similarly, the U.S. Consulate Generals in Hong Kong and Shanghai "have stood out for their use of playful language filled with trendy online [Chinese] expressions" to comment on "hot social and political topics in China" (Tang 2012). Several foreign organizations have voluntarily taken on the task of circumventing China's internal censorship and making this information public (China Tightens Vice on Internet 2006; US in new push to break China internet firewall 2011; Mulvenon 2005). Difficulties associated with software, proxy servers, or VPNs were noted earlier, yet China must continually combat these to maintain control. Both RFA and VOA have been developing alternative methods for getting their radio broadcast into Xinjiang and Tibet, including republishing of reports without identifying signatures, using "SMS to send short messages with proxy information, a peer-to-peer system to distribute content, and Instant Messaging (IM) exchanges, which are less subject to filtering than is email" (Access to Information 2008). Additionally, external IOD is concerned with information available across the Internet; it does not always have to reach China's domestic audience to be damaging. Foreign bloggers using commercially available satellite imagery have compromised Chinese military secrets on numerous occasions. These non-governmental bloggers and private companies have uncovered Chinese aircraft carrier developments, a site used for developing submarine technology, sites used for spy satellite calibration, a training facility used to prepare for a potential conflict with India, and construction of the Wenchang Satellite Launch Center in Hainan (Aroor 2006; Gardner 2008; US Satellite Snaps China's First Aircraft Carrier at Sea 2011; Wolchover 2011). Lastly, copyright infringement of multimedia remains an IO-related Western criticism of China, yet cracking down on this would seem to align with China's interest to reduce Western cultural influence. Ironically, China has had some success in winning copyright infringement cases against the same companies which have on other occasions accused China of copyright infringement or intellectual property theft.

China's use of IO-related lawfare reveals some additional IOD capabilities. China has sought greater intergovernmental control over the Internet, such as attempting to get the Internet Corporation for Assigned Names and Numbers (ICANN) to turn control over to the UN, so they can have greater say in its evolution and reduce U.S. influence (Hsu 2014;

USCC Annual Report 2012). China continues "to promote an Information Security Code of Conduct that advances a state-centric concept of cyberspace and seeks to impose state control of content in cyberspace" (U.S. Department of Defense 2014, 11). Allegations of China violating its WTO obligations claim that China is not only blocking foreign websites but also redirecting users to their equivalent Chinese competitor. Internet "customers in China upon requesting YouTube and Yahoo on their computers received an error message and were redirected to Baidu, China's leading search engine" (Access to Information 2008). The Golden Shield can also slow down access to foreign websites. This might be intentional, or it could simply be the result of the extra processing taking more time; either way it is harmful to business. This has encouraged some foreign companies to relocate their business and servers to a location within China, where traffic will no longer need to go through the big gateways. Not only does this promote economic development in China, but also once in China, these businesses can be subjected to additional laws requiring access to their facilities and products enabling the opportunity for reverse engineering. Chinese law can also enhance their internal IOD by requiring foreign companies to be complicit with China's censorship practices, or reveal the user's personal information as was the case with pro-democracy activists' Li Yibing and Jiang Lijun of Hong Kong, who were using a virtual dead drop (Yahoo Implicated in Third Cyberdissident Trial 2006). Further, in relation to the Golden Shield, "the transfer of financial information from firms like Dow Jones, Reuters, and Bloomberg" must be submitted to Xinhua for government approval before it can be posted online. Relaying financial information to customers quickly is essential for business, and providing this information to Xinhua, a business competitor, places these companies at a significant disadvantage. In addition to advance access to this data, Xinhua has the legal authority "to alter or delete any material they deemed offensive" (Access to Information 2008).

CONCLUSION

Information Operations (IO) influences actions through the orchestrated, non-hacking, control of online information. This encompasses aspects of censorship, media, perception, propaganda, psychology, and public opinion. Relevant literature can be consolidated, given the presence of three core characteristics: the Internet, non-hacking, and attempted sway. This allows a more in-depth examination of IO's full potential, rather than restricted and overlapping views such as the Three Warfares. In comparison to other states, China exercises considerable control over domestic Internet activities and this increases the role of IO. Other states would require hacking to accomplish some of the same actions, which hinders or prohibits their doing so. Further, IO is linked with the other two branches of cyber warfare and plays an important part in China's holistic approach. Information Operations Exploitation revolves around communicating false, misleading, or true messages, intended to cause a desired response. Information Operations Defense guards against undesirable communications by blocking, changing, or countering them. In this sense, IOE is about sending messages, while IOD is about responding to messages. These messages can take multiple forms on the Internet, from government reports and news to social media and memes. Information operations can be used during conflict or peace, and it can be aimed at external or internal audiences.

A large number of government organizations are reportedly responsible for managing Chinese Information Operations (see figure 1.1). For example, the Ministry of Public Security operates the Golden Shield, the Ministry of Industry and Information Technology issues security standards and dealt with the Green Dam Youth Escort, and the State Administration of Press, Publication, Radio, Film and Television released a list of topics prohibited online. Further, the Central Publicity Department has traditionally been responsible for internal IO, while the State Council Information Office has been responsible for external IO. While all of these entities likely provide input or supplementary actions related to IO, the current entities primarily tasked with IO appear to be the General Political Department and the SCIO. A close partnership between these two could represent a more streamlined approach, and it bridges the gap between the State Council and Central Military Commission.

In some ways, IOE is like a popularity contest or political campaign. It can be used to boost perceived strength, soft power, and support for government policies. Alternatively, it can be used to intimidate, sow dissent, and weaken an adversary's soft power. In the 1990s, near the birth of the modern Internet, China began pursuing ways to improve its image through online means; and in the 2000s, government officials began taking IO-related university coursework. In addition to popularity, IOE can be used to subtly convey messages, such as indirectly informing nationalist hackers or protesters when they have the government's support, or when they have gone too far. Similarly, it can be used for sabre-rattling or signalling a likely response to a potential foreign action without committing to an official statement. More unique forms of exploitation include copycat websites, steganography, and virtual dead drops. IOE is most effective when the person conveying the message has a deep understanding of the topic, the transmission medium, and the target audience's language and culture (both physical and online). For this reason, external IOE can benefit from the credence and fluency of foreigners. They can be enlisted directly or influenced through scholarships and special treatment, in order to shape their research or generate a positive view of China, which will carry into their messages. Domestically, China is alleged to have recruited thousands of Internet commentators, in addition to those who conduct IOE voluntarily through shared beliefs and possibly the impact of internal IOE. Tactics for shaping opinions includes the artificial inflation of the votes and view counts of particular articles, comments, and videos. Blurring the line between exploitation and defense is the use of IOE to counter external IOE conducted by foreign states.

A key component of China's Information Operations Defense is censorship, and central to this is the Golden Shield. China blocks or filters content at critical nodes (international gateways), and further content combing is done through a multilevel patchwork of participants and software. These include varying practices used by businesses and Internet service providers, hired censors who manually check content, and self-censorship. This dispersed variety and a lack of transparency give the appearance of unpredictable and sporadic censorship. IO is concerned with perceptions, since these influence action, therefore blanket censorship is not an option. Examples of specific deception tactics employed include drowning out negative messages through an overabundance of preferred messages, and blocking Mandarin language content while allowing the same content in English. The latter makes it difficult for a domestic audience to access, while simultaneously creating a false impression of openness among foreigners. Under the guise of censorship China has also been accused of intentionally slowing, blocking, or redirecting access to foreign websites which are the business competitors of Chinese companies. On the other hand, Chinese IOD faces several problems as the result of interference from foreigners. These include the promotion of ways to circumvent the Golden Shield, backlash over copyright infringement, satellite imagery revealing the location and identity of multiple military installations and developments, and foreign embassies criticizing the government from within China on issues such as pollution. In this regard, China is seeking cyberspace sovereignty. Terrorist groups also pose IOD challenges through their enhanced ability to recruit and target online. The Internet has given individuals powers which were once restricted to large media outlets, and these capabilities include characteristics of anonymity, the near-removal of distance, and a relatively low cost. This exacerbates non-traditional security threats like corruption, criminal organizations, disease, pollution, protests, and natural disasters (particularly relating to building construction and emergency services response) because it allows criticism to spread. Despite some limitations, it is in China's best interest to promote the growth of the Internet as it will boost the economy, improve education, and keep the nation competitive in the twenty-first century. A careful balance, therefore, needs to be maintained between restriction and freedom.

NOTES

1. Some Chinese military analysts use "information operations" as a catch-all term equivalent to cyber warfare or information warfare; however, this book uses a stricter definition (Thomas 2004, 15; Thomas 2007, 29–43, 59–70, 119).

2. These acronyms are not meant to be rigidly defined or delineated here. They are merely useful for discussing and exploring IO, and they have the benefit of resembling the established acronyms used for two subcomponents of CNO.

3. While the government has advocated or allowed such actions, and benefits from them, it is predominantly citizens and businesses who are conducting IO in these examples.

China has a history of reverse engineering websites that become popular and profitable in the Western world; examples include clones of eBay, Facebook, Google, Twitter, Wikipedia, and YouTube being Taobao, Renren (or Sina Weibo), Baidu, Sina Weibo, Baidu Baike (or Hudong), and Youku Tudou, respectively (Denlinger 2010; Goldkorn 2011; Moskvitch 2012).

4. The third branch of cyber warfare, NCW, has the least amount of domestic-foreign overlap, because access and capability are largely confined to states. This is due to NCW involving advanced military hardware that is not connected to the Internet.

5. The opposite of the 50 Cent Party is the Internet Water Army. They are paid by Chinese companies or foreigners to influence opinion in opposition to the CPC.

6. Beyond false statements, "fake news" is a phenomenon which China has had to defend against (Thomas 2009, 115).

FOUR

Net-Centric Warfare

Net-Centric Warfare (NCW)—also called Network-Centric Warfare—is synonymous, or significantly overlaps, with the terms informationization and the Revolution in Military Affairs (RMA). Chinese government and military literature during the decade 2004–2014 predominantly used the term informationization. However, in the absence of a consensus regarding terminology, this book favors NCW. It predates informationization and is more conceptually precise with its inclusion of the word warfare; while the Revolution in Military Affairs, which in turn predates NCW, lacks the implications of *net* or *network* whereby NCW places an emphasis on the role of computer networks and has developed into a more specialized field than the broad concept of RMA. Additional justification for this preference in terminology will be given throughout this chapter. However, more important than the name applied is what these terms collectively denote. They refer to the continuing technologic evolution of military equipment and operations, particularly toward the utilization of computers and computer networks. These computers can be hacked; yet seldom over the Internet, as they tend to be closed networks or are in the process of being transitioned to closed networks to enhance security. NCW is not only hacking into advanced military weapons and C4ISR (Command, Control, Communications, Computers, Intelligence, Surveillance and Reconnaissance), but also the use of those systems. Examples include the use of unmanned aerial vehicles (UAVs) and precision strikes. These might seem far removed from popular conceptions of cyber warfare when used offensively, yet they rely on computers, are susceptible to non-Internet hacking, and help provide an information advantage on the battlefield. As will be demonstrated further in this chapter, this means NCW is distinct from computer network operations (CNO) and information operations (IO). This is an important distinction which is too often neglected in the literature and can lead to misunderstanding. At the same time, NCW shares significant similarities with the other two branches, and this warrants its inclusion under the umbrella term of cyber warfare.

There are additional terms in common usage which connect with NCW, or are components of NCW, yet their relationship is not always made clear. For example, Electronic Warfare (EW) or "dominating the electromagnetic spectrum" is an element of NCW, and it has a large body of literature in its own right. EW includes radio waves, microwaves, infrared, ultraviolet, and X-rays, used for military communication, radar, weaponry, and situational awareness. Just as cyberspace has been de-clared the fifth domain of warfare, on occasion, so too has the electromag-netic spectrum (Bronk 2011; Chase, Engstrom, Cheung, Gunness, Harold, Puska, and Berkowitz 2015; Clifford 2014; International Institute for Stra-tegic Studies 2015; Krekel, Adams, and Bakos 2012; USCC Annual Report 2007; U.S. Department of Defense 2013). This book views them together, under cyber warfare, as the fifth domain of warfare.[1] Another term found in Chinese literature is Integrated Network Electronic Warfare (INEW). Some reports suggest that INEW is a combination of the EW component of NCW and the attack component of CNO, possibly stemming from an increase in the area of responsibility of the 4PLA which traditionally only dealt with EW (Krekel 2009; Krekel, Adams, and Bakos 2012). If that is the case, this pairing could be lost in future shuffles of organization, particu-larly since a holistic approach and pairing of core architectures seems better suited to productivity.[2] Further, additional reports note disagree-ment among Chinese officials themselves as to what INEW denotes, and they make direct comparisons to the term NCW (Salmon 2008; Thomas 2004, 3; Thomas 2005; Wilson 2007). Another term found in Chinese liter-ature, called "system of systems," refers to attempts to link all of the various military networks and weapons platforms together, so they can be used in coordination and share information (International Institute for Strategic Studies 2015). System of systems, including its discussion of the "cognitive, information, and physical domains," is directly taken from NCW and RMA literature which predates China's use of the term. This, therefore, reinforces this book's use of NCW as the preferred branch title. Finally, the term Anti-Access Area Denial (A2/AD) is firmly rooted in the use of NCW, as well as exploiting the weaknesses of an NCW-reliant opponent. In essence, A2/AD literature discusses the use of NCW under the set conditions of two countries at different NCW development stages, with the more advanced state pushing in on the coastal border of the less advanced state. Common components of A2/AD discussion, such as anti-satellite weaponry, jamming, over-the-horizon targeting, radar, ship-de-stroying missiles, and stealth bombers, are all tied to NCW. Some indi-vidual components of A2/AD, like hacking into NIPRNet to delay force deployment, belong to CNO, yet the majority of components frequently discussed in A2/AD articles are firmly connected to NCW. It is the non-

Internet-connected computer networks and proliferation of technology which has made A2/AD distinct from discussions of CNO and traditional war scenarios.

Exploring these terms and concepts presents a structural challenge as there is significant overlap and interweaving. As such, this chapter will begin with a discussion of RMA, because RMA is the oldest of these NCW-related terms and it is the broadest. Next, this chapter will examine NCW specifically, followed by its sub-element EW, and the set conditions of A2/AD. After this base has been established it will be more conducive to show how the preferred Chinese term of informationization is a relabelling of the earlier concepts of RMA and NCW, and why this book prefers to use the term NCW even in the context of Chinese cyber warfare. The informationization subsection will also examine China's current NCW capability and goals. In reality every section in this chapter is about the development path of China, even though the terminology may initially appear to be U.S.-centric.

REVOLUTION IN MILITARY AFFAIRS

The Revolution in Military Affairs is a theory about the future of warfare, and to a lesser degree, a theory about a cyclical pattern of paradigm shifts throughout military history. The current RMA (hereafter simply referred to as RMA[3]) is heavily influenced by the technologic changes of the information age, which may have the potential to yield a dramatic increase in military capability and effectiveness. It is not a single technology which constitutes RMA, but rather the cumulative effect of multiple technologic advancements. Further, while technologic change is the driving force behind it, RMA also refers to changes in doctrine, operational methods, strategies, tactics, training, and organization which make use of those technologic changes. A precursor to the concept of RMA began in Soviet military literature in the 1970s and 1980s. This concept was adopted and expanded on by the defense community of the United States. The Gulf War (1990–1991) is seen as the first example of RMA being implemented in a full-scale conflict. The U.S. military's speed, coordination, precision, low casualty rate, and overall technologic dominance in this war increased interest in the study of RMA domestically and internationally, including in China (Bronk 2011; Chase, Engstrom, Cheung, Gunness, Harold, Puska, and Berkowitz 2015; Metz and Kievit 1995; Thomas 2004, 47–48; Thomas 2007, 71). U.S. technologic superiority and interest in RMA has continued, most notably with the Kosovo War (1998), the War in Afghanistan (2001–2014), and the Iraq War (2003–2011). In 2004 Vice President Richard Cheney stated:

> With less than half of the ground forces and two-thirds of the military aircraft used 12 years ago in Desert Storm, we have achieved a far more

difficult objective. . . . In Desert Storm, it usually took up to two days for target planners to get a photo of a target, confirm its coordinates, plan the mission, and deliver it to the bomber crew. Now we have near real-time imaging of targets with photos and coordinates transmitted by e-mail to aircraft already in flight. In Desert Storm, battalion, brigade, and division commanders had to rely on maps, grease pencils, and radio reports to track the movements of our forces. Today, our commanders have a real-time display of our armed forces on their computer screen. (Raduege 2004)

Key changes associated with RMA include precision strikes and improved C4ISR, as well as a push toward streamlined and rapidly deployable forces such as special operations forces. However, the latter may reflect the mission needs of recent conflicts; this could change in the event of conflict between technologically advanced near-peer competitors. In either case RMA may represent a shift from centers of gravity,[4] like massed forces, toward a "more sophisticated notion of interlinked systems" (Metz and Kievit 1995). For example, dispersed and interchangeable assets could provide increased efficiency and redundancy, comparable to peer-to-peer software and cloud computing. RMAs are composed of multiple minor and major technological and organizational advancements. Not every minor change must be adopted in order to remain a top military competitor. Indeed, perusing every minor advancement can overstretch resources, reduce efficiency, and create weaknesses. However, major changes in the conduct of war, revolutionary components which will endure for decades or centuries, cannot be ignored without falling behind the competition.

A commonly held belief among RMA analysts is that the Revolution in Military Affairs will have at least two stages. The first stage is comprised of components which have become increasingly familiar, such as improved C4ISR, global positioning systems, information dominance, interoperability, jointness, missile defense, precision strikes, and stealth. These in turn create characteristics which are associated with RMA. For example, guided munitions or smart bombs have reduced casualties and collateral damage. Similarly, improved situational awareness, including the ability to identify friendly or foe positions on a battlefield, has reduced friendly casualties while increasing the tempo and lethality of operations. There is no definitive source which outlines which technologies belong to the first or second stage, yet the second stage is generally believed to still be in its infancy and will result in a more fundamental change in the conduct of warfare than the first. Elements which have been associated with the second stage of RMA include attempts to manipulate the ionosphere, like the High Frequency Active Auroral Research Program (HAARP); biotechnology and mind-machine interfaces; clean bombs, including theoretical antimatter weapons which would reduce nuclear fallout; lasers, such as the field-tested Laser Weapon System; nanotechnology; robotics; the use of holographic imagery; and the weaponization of outer space. One subcategory of robotics is powered exoskeletons, which could allow soldiers to carry heavier loads and wield larger weapons while simultaneously increasing their hiking endurance. Further, these exoskeletons would aid logistical crew by improving their ability to load heavy armaments, equipment, and supplies. Non-lethality is also a frequently discussed feature of RMA, which could include the use of adhesives and foams, or directed energy weapons such as acoustic, electromagnetic pulse (EMP), and microwave weapons. Acoustic weapons have already been used to combat piracy in the Gulf of Aden, and the United States temporarily deployed a microwave weapon, the Active Denial System, in the War in Afghanistan (Evers 2005; Pike 2011). Additional components which have been ascribed to the second phase of RMA include increased soldier protection, such as nanotechnology-enhanced body armor; "instantaneous global communications" including mobile phones; and "the pervasiveness of the electronic media," such as a soldier being able to see the battlefield and troop positions in realtime on a tablet device or heads-up display (Metz and Kievit 1995).

As previously stated, RMA encompasses multiple technologic advancements which create an overall picture, yet the advancements which will truly stand out might only be clearly known in retrospect. For this reason, not all authors include the same items, and it is the broad concept which is more important. This is exemplified by authors Sean J. A. Edwards and Martin Libicki of the Rand Corporation, who have written about BattleSwarm tactics and "fire ant warfare" in which thousands of small machines could swarm a target (Edwards 2004; Libicki 1994). Other authors discuss swarm tactics, but in a different context. For example, soldiers could present problems in a chat room environment where globally dispersed experts could collaborate on solutions, and information can be gathered from a decentralized source, reminiscent of BitTorrent.[5] Further, networked troops can operate in smaller units with loose formation, because they have greater awareness of friendly and foe positions; if a unit comes under fire or locates a valuable target, the other units can swarm to their position. These dispersed, yet connected, small units can cover more ground and are more difficult for an enemy to target. RMA analysts might dispute the feasibility or originality of particular systems, yet the broad elements of nanotechnology, networked systems, robotics, and semi-autonomous systems will likely remain an element of RMA. Beyond the two phases of RMA development, there is also a cyclical pattern to RMAs which begins at relative equilibrium. This equilibrium is disrupted by the implementation of new technologies. Once this collection of various technologies reaches a critical mass, there are consolidation and response periods before returning to stasis. It is uncertain which technologies or plans of implementation will survive the consolidation and response periods. One of the difficulties of implementing RMA is attempting to decide which elements to adopt. Perusing all of them could cause bankruptcy or confusion, yet there are advantages to early adoption of elements that are able to endure successfully. Some analysts advocate focusing modernization efforts on the second, more fundamental, stage and advancing at a more moderate and cautious pace.

States attempting to utilize RMA face a number of challenges, and these challenges remain present in the NCW and informationization iterations discussed later in this chapter. RMA is most effective against a mid-level opponent, one which has a "rigid centralized decision making" process and relies on "limited range, easy-to-detect weapons platforms such as tanks, conventional artillery, and manned aircraft" (Metz and Kievit 1995). Peer-level adversaries possess comparable abilities, which can offset the dramatic advantage seen against mid-level opponents; they also understand common RMA-associated weaknesses and can bolster their defenses or create countermeasures accordingly. At the other end of the spectrum, low-level opponents, such as terrorists, tend toward asymmetric and guerrilla warfare which can exploit the weaknesses of an RMA-reliant force. They are often widely dispersed, operate in cells, blend in with the civilian population, and "emit a limited electronic signature," all of which erode opportunities for targeting and make a clear-cut victory difficult to determine (Metz and Kievit 1995). Similar to computer network attack and computer network defense, a terrorist organization might only need to identify one weak point for an attack to have the desired impact, whereas the defender must consider all possible exploits. The complex nature of advanced technology and its propensity toward networks could present a greater attack surface for adversaries. Further, over-reliance on these technologies could cripple an RMA force in the event of a system failure. Discontinuing the *old ways* of operating could significantly hinder their reintroduction should they once again be deemed necessary, yet maintaining both is costly and inefficient. Additionally, if systems become too complicated or the bugs are not fully worked out, they can become counterproductive. There is also a high financial cost associated with advanced technology and attempting to completely restructure how forces are organized and operate. The technology itself is rapidly changing, which makes it hard to keep pace. Additionally, the more that states invest in RMA, the more adversaries will invest in countermeasures, and this exacerbates the cost factor. One example is the use of a multiple independently targetable re-entry vehicle (MIRV) to counter ballistic missile defense; another is the cost of an ASAT or laser versus the cost of launching and operating a satellite.

Despite reduced casualties on the side of the RMA-advanced force, and claims of reduced collateral damage, many of the elements of RMA, such as non-lethality, precision strikes, and robotics carry ethical concerns and new challenges. Precision air strikes, for example, have become the accepted norm for advanced militaries. A return to carpet bombing would cause a severe backlash, and yet precision strikes themselves are a regular source of international outcry. Without a human presence on the ground, it can be difficult to disprove claims of civilian casualties or conduct battle damage assessment, and global communications have dramatically increased the ability of those attacked to transmit opposing messages. While heightened scrutiny over collateral damage is a noble cause, it also increases the expectations and financial cost of warfare. The Israeli Defence Force has gone as far as dropping leaflets, telephoning neighborhood residents, and using radio to warn of their own impending strikes. They have also adopting a "roof knocking" tactic in which a low-yield explosive is detonated on the roof of a targeted building. This rattles the building and acts as a final warning to evacuate, moments before the full-strength missile strike. Additionally, the use of drones has raised criticism of the dehumanizing effects of RMA. Drone operators may be located on the opposite side of the planet from their drone and go about a normal life outside of work hours. Furthermore, their use of a joystick and computer screen gives it the appearance of a video game, and critics feel this is causing a disassociation. However, it can be argued that manned aircraft dating back to World War I have always been distanced from their targets, and many other weapon systems strike from a distance. It is unclear at what point a strike would be considered ethical, if ever, under conditions of war, yet the criticism remains. (Drone use will be discussed further in the following subsections of this chapter.)

THE DEVELOPMENT OF NET-CENTRIC WARFARE

Net-Centric Warfare is a theory about warfare in the information age that was promulgated by the U.S. Department of Defense and U.S. armed forces from the 1990s (Alberts, Garstka, and Stein 2000). In essence, NCW is the application of information and communication technology (ICT) to provide a battlefield advantage. More specifically, NCW seeks to integrate "command," "sensors," and "shooters" into a shared network (discussed below). This can enhance coordination, information sharing, lethality, precision, situational awareness, speed of command, speed of communication, and tempo of operations. Further, it allows for distributed weapon assignments, engagement from greater distances, geographically dispersed collaboration, an increased range of missions capable of being undertaken, rapid target assessment, a reduction of combat unit sizes (which in turn allows them to travel lighter and faster), and superior decision making (Electronic Warfare 2007; NCW Roadmap 2007; Network-Centric Warfare: Creating a Decisive Warfighting Advantage 2003). In turn, all of these can improve overall combat effectiveness and efficiency. Net-Centric Warfare is not only known as Network-Centric Warfare (as noted above), but is largely interchangeable with the terms

Network Centric Operations and Future Combat Systems; NCW literature often notes the goal of obtaining an information advantage, information dominance, or information superiority (Raduege 2004; Wilson 2007).
Use of these information terms can lead to misunderstanding, particularly in mass media, such as falsely assuming these networks are Internet-
connected. ICT enables NCW, but these technologies need not be connected to the Internet as that would reduce their security. "Information
superiority" becomes clearer when considering the range of items which
can be viewed as "information" in warfare (Thomas 2004, 39). For example, the locations of friendly, foe, and neutral people on a battlefield is
valuable information and it can be transmitted as data. The Internet has
become so pervasive that it has become common to associate all computers as having an Internet connection (a trend which has expanded to
mobile phones as well). Yet computers existed before the Internet, and
modern weapons and weapons platforms, such as sniper rifles or the F-35
aircraft, utilize computers in their operation. Similarly, hacking has become commonly associated with the Internet, yet these isolated computers and non-Internet networks can also be hacked. U.S. military officials
have warned that deployed aircraft, ground systems, missiles, satellites,
and ships are all vulnerable to hacking, and China has been specifically
cited as perusing this capability (USCC Annual Report 2012).

The Iraq War (2003–2011) is seen as a momentous occasion for the
development of NCW, just as the Gulf War (1990–1991) was for RMA.
U.S. troop levels were approximately 30 percent less in the first year of
the Iraq War compared to the Gulf War, yet the Iraq War consumed 96
percent more bandwidth, suggesting the possibility that Washington was
able to accomplish its goals with fewer troops due to an information
advantage (Raduege 2004). This was despite the fact that NCW in the Iraq
War was still deemed "transitional rather than transformational"; many
of the NCW systems were being field-tested on a trial basis and not all
units had the same level of networked equipment (Wilson 2007). As with
RMA, NCW is not only a collection of new technologies but also entails
their optimal use through changes in doctrine, training, and organization.
Much of what constitutes NCW will require further implementation, testing, and debate before it is fully developed. Many computerized military
systems are already networked, such as an unmanned combat aerial vehicle (UCAV) being controlled by an operator who is far removed from
the battlefield. The operator can receive orders from command in additional locations, or receive input from soldiers on or near the field of
battle. Using NCW terminology, the operator is the "shooter" and the
UCAV is the "sensor" and weapon platform. Soldiers on the field providing additional input (battlespace awareness) are also "sensors," and the
commander giving orders or determining tasking is the "command."
Each of these computerized inputs are a node within the network. Some
NCW authors also discuss physical, information, and cognitive domains.

The physical domain is the tangible environment, such as the sensors, platforms, and shooters. Information gathered by the sensors is transmitted as data through the information domain and processed by human recipients in the cognitive domain (the human brain) (Ahvenainen 2003; Alberts, Garstka, and Stein 2000; Cyberspace Operations 2013; Information Operations 2012). As another illustrative example of existing computerized information networks,[6] a strike aircraft can communicate with command, receive supplemental radar information from Airborne Early Warning and Control aircraft, and utilize global positioning system (GPS) satellite or ground-assisted laser guidance[7] targeting.

One of the proposed goals of NCW is to link all of these existing networks under one unified network, creating a "system of systems," and linking all branches of the armed services together to create a synchronization like RMA's interoperability and jointness. Efforts to create an overarching network are still in development. One possibility being pursued by the United States is the Global Information Grid (GIG), which seeks to interconnect all sensors, shooters, weapons platforms, and command and control. The GIG initiative seeks to coalesce the U.S. Air Force's C2 Constellation, the Army's LandWarNet, and the Navy and Marine Corps' FORCEnet programs (Wilson 2007). A supplemental goal of this interconnection is to create an environment where information is "pulled" rather than "pushed." Traditionally information is pushed down the chain of command to those who need it; this is sometimes instigated by a request from the lower levels, or the higher levels attempt to anticipate the needs of the lower levels. Under a unified network, all information could theoretically be accessed (pulled) on demand by "authenticated users within a given community," thereby reducing the amount of time it takes for dissemination and promoting greater depth of "information sharing and collaboration" (Wilson 2007; Raduege 2004).[8] This partially non-linear decentralized approach to command and control would also give more freedom to individuals and tactical units.

The inevitability, and some of the risks, of networked systems can be illustrated further using the above mentioned unmanned aerial vehicle networks. While the use of these "drones" (as UAVs and UCAVs are commonly referred to) remains contentious, their numbers and use continue to rise and additional markets are opening up (Fritz 2013a). The International Institute for Strategic Studies identified 807 drones in active service around the world, not including any that may be in use by China, Turkey, and Russia (Rogers 2012). As of 2013, some estimates placed the number of enemy combatants killed by U.S. drones at 4,700, and this figure has likely risen following American-led intervention in Iraq and Syria against the Islamic State of Iraq and the Levant which began in 2014 (Knox 2013; Dwyer 2013). According to an article in the *Wall Street Journal*, the U.S. Air Force has "staked its future on unmanned aerial vehicles" with drones accounting for 36 percent of all new planes in its proposed 2010 budget and plans "to buy as many as 375 Reapers" (Gorman, Dreazen, and Cole 2009). The Global Hawk reconnaissance UAV, and Predator and Reaper UCAVs, rely on satellites to communicate globally with command and control (Norris 2010, 230–32; Wong and Fergusson 2010, 59). UAVs have already come to the attention of hackers. Insurgents in Iraq and Afghanistan used a 26 USD off-the-shelf software program called SkyGrabber, from Russian company SkySoftware, to capture unencrypted Predator video feeds. SkyGrabber accesses data being broadcast by satellites, allowing users to tune into different data streams comparable to tunning in broadcast radio stations (Ward 2009). This did not allow the insurgents to control or disrupt the UAVs; it only permitted them to eavesdrop on the signals being sent. This, however, bore a number of alarming implications. First, of all types of satellite operators the military is thought to implement the highest level of security measures, and yet encryption was not being used. Once this particular weakness was exposed it had to be fixed, which added cost to the program, took time to implement, and reduced the efficiency of the equipment. Secondly, the information gained by insurgents might have revealed the areas that were under surveillance and patterns of drone use, providing them with information they could use to avoid detection or set up an ambush (Gorman, Dreazen, and Cole 2009). Further, while SkyGrabber did not allow hackers to take control of the UAVs, control and other hacking options remain theoretically possible. Under asymmetric warfare, exploiting weaknesses in drones and an opponent's reliance on them could be used as part of a larger "blinding campaign" (Krepinevich 2010; Tol, Gunzinger, Krepinevich, and Thomas 2010). As Captain J. W. Rooker of the U.S. Marines states, "If an enemy can disable or destroy the satellites on which these systems depend, or hack, jam, or spoof them, he will have effectively gouged out our eyes" (Rooker 2008).

Net-Centric Warfare not only utilizes networked systems but also denies an adversary's use of such systems while protecting one's own. The increased range of conducting warfare, as well as the new targets and types of targets these systems present, has expanded the battlefield (Thomas 2004, 28). Further, the reduced size of deployed forces, data hubs, and critical network nodes could become concentrated high-value targets. Despite NCW's goals of increasing information sharing through networked systems, not all information from every system is meant to be shared with each node. If this were to occur, it could result in information overload, increased bandwidth demands, reduced speed, and security or command and control concerns. The degree to which nodes are to be connected must be determined by examining the roles and responsibilities of each entity. This relates to the above observation that NCW represents more than advancements in ICT; it also concerns changes in doctrine, training, and organization. The goal is to create sets of interconnections which can enhance situational awareness and decision making through greater availability of information. Commanders must be able to "allocate, assign, and employ assets and then modify these allocations, assignments, and employments as awareness of the situation changes" (Alberts, Garstka, and Stein 2000). In addition to the swarm tactics discussed in RMA, sensors can also be tasked to focus on an area of interest or to fill in gaps in the network due to deficiencies in capability or the destruction of platforms. This introduces an element of redundancy and self-healing into the network. Net-Centric Warfare is a holistic evolution of traditional military equipment, doctrine, training, and organization. The wide range of systems being incorporated into NCW, along with the technical aspects, are beyond the scope of this book. However, some recurring concepts are examined in the following paragraphs to provide a deeper understanding of NCW. These include architectural aspects of NCW, improved targeting capability, software interfaces, improved logistics, and problems or concerns with implementing NCW.

Exploring the architecture of an NCW system of systems is complicated by the wide variety of systems which states are attempting to incorporate, the classified nature of the topic, and the use and modification of commercial ICT products to suit closed military networks. Military radio (voice communications) alone has a long and complex history dealing with the delineation of frequency bands, interoperability of equipment, cryptography, and ensuring secure transmissions. NCW is not only attempting to incorporate this, but also all communications, command and control, sensors, shooters, and platforms, across all military branches. Each of these are large and diverse categories. One need only consider how many different types of weapons platforms there are. Within these elements several systems have already been formed, such as multiple air defense, intelligence, navigation, radar, reconnaissance, surveillance, targeting, and weather-forecasting systems. These systems can be components of larger systems, and each is comprised of varying technical specifications, hardware, software, and means of connectivity. Some networking efforts, such as creating tactical data links[9] between nodes, have undergone development in an isolated fashion by being restricted to specific programs or service branches.[10] This adds to the difficulty of creating a comprehensive system, as well as the difficulty of researching this topic. Tying all of these systems together while maintaining security, reliability, and speed is a significant undertaking. However, the limitations of science and technology may have forced some initial conformity, because there are only a select number of ways to transmit and receive data effectively and efficiently. This conformity will be helpful in integrating these systems; however, it also has the unintended consequence of furthering the misconception of Internet connectivity. NCW appears to be built on common elements found within Internet architecture, such as fiber-optic cables, the use of Internet Protocol (IP), Internet-like routers, satellites which rely on radio and microwave transmissions, and line-of-sight towers (comparable to those used for mobile phones) to provide a high-bandwidth communication backbone (Wilson 2007; Raduege 2004). In some cases, these are modified for enhanced security, and NCW does use some methods which are uncommon within Internet architecture, like infrared, or lasers used for concentrated burst transmissions. Given NCW developmental status, some networks which will ultimately become a part of this system of systems may currently possess limited Internet connectivity. The end goal is to remove these connections and rely on closed networks to enhance security. For example, China has built "dedicated fiber-optic command and control networks" which are air-gapped from the Internet (Krepinevich 2010). In addition to being separate from the Internet, they are "hardened and buried" to reduce the risk of NCW-style hacking, EMP, or physical destruction (Tol, Gunzinger, Krepinevich, and Thomas 2010). China has also been developing "new satellites for establishing a unique [military-only] GPS network" (Wilson 2007). These are in addition to the previously mentioned closed network, which China is pursuing for classified information (Thomas 2004, 10–11). Due to commonalities in hardware and software, some of the malware discussed in the CNO chapter could be effective under NCW, but the infection, communication, and exfiltrated data would not transmit directly over the Internet.

Despite the enhanced security offered by closed networks, illicit access can still be gained through wireless (electromagnetic) connections or removable data storage, such as USB thumb drives, memory cards, or compact discs. In 2012, for example, China was accused of using removable data storage to access a closed network at India's Eastern Naval Command (USCC Annual Report 2012). Additionally, intruders can access these networks through the same fiber optics and wireless connections that legitimate users employ, if the networks are exploited or credentials are falsified. However, these networks may be more difficult to access than Internet networks because peripheral entry cannot be obtained through a commercial Internet service provider (ISP). Further, their workings and technical specifications are likely to be classified and the networks are likely to be operated under high levels of security, including encryption, monitoring, and geographic proximity requirements. The distinction between Net-Centric Warfare and Computer Network Operations is complicated by some overlapping gray area and combined effects, such as the previously discussed accusation that China eavesdropped on conference calls pertaining to classified F-35 designs. In that case, CNO allegedly allowed China to obtain information on the F-35's software and "secure communications and antenna systems" — information which could be used separately to exploit closed networks under NCW-style hacking or to degrade the effectiveness of NCW-enabled equipment (Reed 2012). As another example of a gray area, the U.S. military has sought the development of USB flash drives or similar devices which can act as automated forensic tools. These tools can quickly extract intelligence from enemy combatant computers on-site, such as during the raid of Osama Bin Laden's compound in Pakistan and raids of the suspected homes of high-value targets within Iraq (Cox 2014; Goodin 2009). Fast retrieval limits the amount of time soldiers must remain in a dangerous position and it eliminates the need to transport heavy or cumbersome computers back to base. The computers which are searched may or may not be Internet enabled, yet such devices appear more in line with NCW than CNO because the data retrieval is occurring on a battlefield at its place of origin rather than over the Internet.

The networking of sensors, shooters, and command has led to an increase in the tempo of operations and the effectiveness of targeting. Illustrative of this is communication with missiles that allows their trajectory to be changed in-flight in the event that a target has moved, the target has already been destroyed, or commanders decide to abort the attack (Wilson 2007). It also permits a missile to be fired by one platform at great distance, such as a ship, and then guided to its target by another platform which is closer to the target, such as an aircraft, in a move similar to passing a ball downfield in sports (Macri 2015). This can extend the range and accuracy of missiles and reduce shooter casualties as drones could guide them in the final stage. It also reduces the amount of ordnance these drones must carry, which enhances stealth, reduces fuel consumption, and increases the amount of time they can loiter over potential targets. Bombers, too, have greater flexibility through networked data links and satellite-guided (smart) bombs, as improved targeting data or changes in the situational assessment may become available after takeoff (Alberts, Garstka, and Stein 2000). Separate strikes which had not yet occurred at the time of their departure can be taken into account and corrective measures incorporated. Further, the time it takes analysts and commanders to identify targets through satellite or aerial imagery and issue orders against those targets has been reduced through the use of computer networks. In the Gulf War this took up to four days, but by the time of the Iraq War this was reduced to around 45 minutes (Wilson 2007). The U.S. "shock and awe" campaign was not meant purely as a measure of destruction. It also referred to the speed and accuracy with which a large number of critical targets were struck. Increasing the speed of decision making and actions is essential as there is limited time to "detect, track, classify, and engage targets" (Alberts, Garstka, and Stein 2000). Improved sensor technology has increased the efficiency and effectiveness of strikes. This is in part due to advancements in technology and control of the electromagnetic spectrum (discussed in detail below); however, it is also due to computerized networks allowing multiple sensor data to be rapidly shared and utilized. Radar and thermal cameras are able to detect movement and heat signatures in rain or darkness, and aircraft equipped with smart bombs have the ability to conduct close air support during the poor visibility of sandstorms without suffering friendly fire (Wilson 2007). Some networks, such as the previously noted Advanced Tactical Targeting Technology and Cooperative Engagement Capability, allow multiple aircraft and ships to combine their individual sensor pictures (including radar) to create a real-time composite image and air defense network. This can allow ships to shoot at targets from great distances, targets they could not see through their own sensors alone, such as incoming anti-ship missiles. During the Iraq War, some Bradley Fighting Vehicles and M1 Abrams tanks "replaced paper maps and routine reporting by radio voice communication" with digital data (Wilson 2007). They were capable of pinpointing positions during adverse weather conditions, and they could be tracked and monitored by commanders overseas. Additionally, some ground units field-tested eyepiece displays and tablet-like devices for enhanced battlespace awareness, as well as lightweight short-range drones and unmanned ground vehicles in order to see what was around corners or over hills (Raduege 2004).

Battlespace awareness and the effectiveness and efficiency of networked military systems can be enhanced through computer aids and user interface software. Linking together multiple systems, such as the previously mentioned radar, can create a larger picture, yet traversing that picture as a unified whole and making useful sense of it in a timely manner can be problematic. This is particularly difficult when considering the large array of systems which NCW seeks to interconnect. Ultimately, NCW seeks to have a three-dimensional map of the battlefield comparable to a modern video game. A user could zoom in on a single slice of the battle, such as a single firefight or the status of a single vehicle, and zoom out to see the entire theater of operations, including overall statistics. At any given scale, various components on the map and available types of information could be added or subtracted from display. This map would need minimal lag time and distortions to be effective. Some templates or models of recurring elements, known as common operation pictures (COPs), could help in this programming. According to Alberts, Garstka, and Stein's *Network Centric Warfare: Developing and Leveraging Information Superiority*, a standard COP might include the following four elements:

1. Location (current positions, rate of movement, and predicted future positions);
2. Status (readiness postures including combat capability, whether or not in contact, logistics sustainability, and so forth);
3. Available courses of action . . . [ability to call for air support or reinforcement, assessment of enemy capabilities, and the predicted enemy courses of action];

4. The environment (including current and predicted weather conditions, the predicted effect of weather on planned operations and enemy options, and terrain features such as trafficability, canopy, sight lines, and sea conditions).

All of this information would be continually updated to include movements, changes, and battle damage assessment. A simplified version of a COP has become common on commercial airlines—individual passengers can view a map which displays their current altitude, heading, location (with an airplane-shaped icon), and speed; it automatically zooms in and out, showing where they have been and where they are going, as well as displaying additional information such as which parts of the planet are in sunlight or darkness; and some planes also switch to a live video feed from a nose-mounted camera during landings. Military interfaces aim to have much more detail and interactivity than this, but it illustrates the technologic feasibility of such an undertaking. The capability and usefulness of automated alerts, support, and helpful information might also be easier to accept when considering all of the alerts and pop-ups on a modern computer operating system, or even the "check engine" light on an automobile or the autopilot on an airplane. Similar to the planning required to create COPs—decision aids,[11] "doctrinal models," and rule sets can be made to help users (soldiers) make decisions in minimal time which have the highest probability of success (Alberts, Garstka, and Stein 2000). These are predetermined responses to set conditions, which a user can choose to employ, or in some cases can be set as an automated or semi-automated response. Some systems, such as Northrup Grumman's carrier-based drone, the X47-B, and fly-by-wire technology already rely on some autonomous programming (Skillings 2012; Weinberger 2012). Net-Centric Warfare does not entail relinquishing control to computerized systems. Doing so would likely create a public backlash, but there are some instances, such as logistical support, where automated systems would be beneficial and cause minimal ethical concerns. Since they help to achieve stated goals, the amount of automated systems in use seems likely to increase as these systems develop. Similar to RMA, robotics, nanotechnology, and biotechnology are a part of NCW which remains in the early stages of development.

Some of NCW's early concepts were derived by examining how advancements in ICT had improved operations in the business world, such as improving the ability to monitor competitor prices, customer satisfaction, inventory, production, shipping, and situational awareness, including machine malfunctions or low stock. Orders can be taken and sent electronically, price changes determined by a head office can be set to occur automatically within stores, the bestselling products and trends can be identified and analyzed, and customers can be sent to sister stores if an item is sold out. As in NCW, these increase the speed of command and operations, allow operations and projects to be coordinated from great distances, allow well-informed decisions, free up time and resources, allow smaller tactical units to conduct operations effectively, and limit a competitor's plans and options. In both the commercial sector and networked warfare, the "relevance, accuracy, and timeliness" of information are crucial to success (Alberts, Garstka, and Stein 2000). Increasing efficiency and allowing portions of the team to participate from great distances reduces the number of personnel needed in the theater of operations. This creates a smaller battlefield footprint allowing forces to travel "lighter and faster," including during the initial deployment phase (Wilson 2007). For example, fewer soldiers and personnel mean less equipment, food, and supplies that need to be shipped. Once NCW infrastructure has been fully developed, this reduction in size could translate to a lower cost as the movement of information is cheaper than that of personnel and equipment. Efficiency itself provides cost savings, and NCW provides multiple avenues for increasing efficiency and speed. For example, teleconferences for planning, virtual collaboration for problem solving, automated delivery orders for when supplies become low, and electronic manifests for shipping. Previously supplies and soldiers had to be manually accounted for on clipboard and paper before departure; this time expenditure has been significantly reduced by using scanners and wireless identification. In the theater of operation, data links between platforms can also greatly enhance logistics. Aircraft or vehicles can relay their status—whether it be items that need maintenance, fuel requirements, or armament levels—to base or an aircraft carrier, so that these items and associated teams can be prepared to respond appropriately. Ordnance or replacement parts, for example, might need to be moved to a staging area, which can be a timely process. Additionally, knowing the status of damage, fuel consumption, and armaments can help commanders prioritize a queue for resupply or reassign platforms which are capable of responding to further missions without a return to base. Other activities which are removed from fighting, yet integral to mission success, are also enhanced by ICT.[12] Troops can communicate with family overseas on a more frequent basis through e-mail, satellite phone, or video conference, which helps to increase morale. Some troops and commanders who operate from great distances, such as drone operators, can maintain their normal off-duty activities in their home country. Lastly, deployed troops no longer have to take leave for supplemental education and training programs as these can be conducted electronically.

Implementing NCW presents multiple problems and concerns. Complex electronic systems are subject to technical difficulties, interruptions to service, and even cascading failures. If forces become too reliant on using information networks and are faced with a disruption to its service, they might no longer have the support or experience of previous warfare methods. Maintaining backup systems and training would increase cost and occupational demands. Further, if hackers are able to exploit NCW systems, they could introduce false data to send troops and missiles to the wrong targets. These problems are also relevant to essential logistics support, particularly in the buildup to war and deployment phases, which are increasingly reliant on information networks (Mulvenon 1999). Information overload, or signal-to-noise ratio, is another concern. Individuals could be presented with so much information that the useful parts get lost amid the unessential information, or an overabundance of information creates a distraction and reduces response time. As with computer network defense, there is concern that the software these systems use could be compromised by embedded backdoors (Thomas 2004, 45). One way to overcome this is to require all software to be produced domestically and under strict security procedures; though, this would escalate the cost of an already expensive transformation. Increasing security is often accompanied by a rise in cost and a decrease in speed. The implementation of encryption, which slows down speed, demonstrates this; so does the costly hardening (shielding) of NCW equipment to protect against electromagnetic pulse attacks or natural phenomena like solar flares. The use of commercial entities and Internet connections in developing NCW capabilities saves money at the cost of reduced security; although cost saving does not appear to have been the motive. Rather, development of an NCW information backbone has yet (in 2015) to meet demand. This does however present an interim problem. During the first half of the Iraq War, approximately 80 percent of U.S. satellite-enabled bandwidth was provided by commercial entities, placing them in greater danger of computer network attack and exploitation (Wilson 2007). Even when robust closed networks are operational, bandwidth can pose problems as the volume of data that system of systems will generate is potentially enormous. Beyond disruptions to service, a lack of timeliness can be a serious concern. For example, in 2003 a U.S. Army battalion was "surprised by a large force of Iraqi tanks and troops," because their information network was "unable to update enemy information in databases quickly enough to keep front line units accurately informed" (Wilson 2007). The challenge of creating interoperability among multiple developing systems is exacerbated by a need to cooperate with allies or engage in military operations other than war (MOOTW). Partners will want a free-flow of information, while maintaining some restrictions for security purposes, and the ability to add or remove access as missions end or alliances change.

ELECTRONIC WARFARE

Electronic Warfare (EW) is the use of the electromagnetic spectrum to enhance combat capability. As stated in this chapter's introduction, some authors view EW as a distinct domain of warfare; however, this book deems EW to be a component of NCW. The electromagnetic spectrum (EMS) includes X-rays, ultraviolet, visible, infrared, microwaves, and radio, all of which are intertwined with NCW systems.[13] For example, sensors use the visible, infrared, and radio (radar) portions of the EMS to gather targeting information. This information is often transmitted as data to commanders and shooters via radio waves, and in some cases shooters use the EMS itself as a weapon, such as lasers and microwaves. Electronic warfare literature notes that EW has been used for more than a century, primarily as a means of communication; however, "new technologies" are ushering in an EW military "revolution" (Electronic Warfare in Operations 2009). Even military radio communications, which must transmit through an increasingly crowded electromagnetic environment[14] while maintaining secrecy and reliability, have become so complex that computers are required in their operation. These and other recurring statements mirror the Revolution in Military Affairs and Net-Centric Warfare. Common topics of EW discussion include the ability to act from great distances, C4ISR, integrated air defense, precision and non-lethal weapons, robotics, and situational awareness. Further, EW theorists discuss how the spread of commercial ICT products, including a growing trend toward wireless technologies, has given less technologically advanced militaries and non-state actors the ability to compete in this arena; this correlates with NCW, as well as the upcoming section on anti-access area denial. Additionally, EW seeks to enhance the capabilities of sensors, shooters, and command, and the primary means of transmitting information within NCW is wirelessly through the EMS.[15] While it is explicitly addressed in EW, all of this chapter's subsections involve military attempts to control or "dominate the electromagnetic spectrum" (Electronic Warfare in Operations 2009). In other words, armed forces are developing ways to "exploit the opportunities and vulnerabilities that are inherent in the physics" of the electromagnetic environment (Electronic Warfare 2007).

Electronic warfare is composed of three components: electronic attack (EA), electronic defense (ED), and electronic surveillance (ES). Some EW literature uses slightly different terms for these three components; however, the ones chosen here are the same as those used by the North Atlantic Treaty Organization. They also have an added benefit of elegance, since they resemble CNO's attack, defense, and exploitation components. Some EW literature uses the terms electronic protection (EP) in place of electronic defense, and electronic warfare support (also ES), instead of electronic surveillance (Electronic Warfare 1999; Electronic Warfare 2007; Electronic Warfare in Operations 2009). The definitions of ED and EP are interchangeable; however, surveillance and warfare support have different connotations. This book uses ES to refer to both surveillance and warfare support collectively. They already share the same acronym, and only together do they encompass all of the remaining elements of EW, which do not fall under attack or defense. It is also important to note that there are some discrepancies between authors as to which component specific activities belong. Some items which are described below under ED, for example, are described by others as "defensive electronic attack" under the category of EA (Electronic Warfare in Operations 2009). Jamming transmissions is illustrative in that it could be considered defensive or offensive; it might prevent enemy precision weapons from being guided and triggered, or it might knock out enemy air defenses. In the absence of consensus on some categorization details, the following discussion is organized in what was deemed to be the most logical way for conveying the larger picture as it relates to NCW. Regardless of subcategory, the following items are universally recognized as belonging to EW as a whole, and they utilize the electromagnetic spectrum. The goal of EW is to obtain an advantage through the use of the EMS, to maintain unimpeded access to the EMS, and to deny an adversary's ability to do the same. EW supports military activities through various types of "detection, denial, deception, disruption, degradation, protection, and destruction" (Electronic Warfare 2007).

Electronic attack (EA) involves using the electromagnetic spectrum to degrade, destroy, or neutralize an opponent's equipment, facilities, people, platforms, and sensors. EA includes the use of directed energy weapons, such as dazzler, electromagnetic pulse, laser, and microwave weapons. Dazzlers can be used to disrupt the optics of surveillance satellites, or small-scale rifle-mounted dazzlers can be used to blind temporarily or confuse enemy combatants during a raid.[16] Laser and microwave weapons include the previously mentioned U.S. Navy's Laser Weapon System and the Active Denial (microwave) System. Anti-radiation missiles which lock on to sources of radiated energy, such as enemy radio emissions, are also EA weapons. Infrared homing missiles, or heat seekers, are another type of anti-radiation missile. Jamming, which will be discussed further below in relation to satellites, can degrade or prevent the use of enemy infrared, laser, and radar guidance systems. Additionally, jamming can limit or prevent enemy communications and situational awareness sensors, thus degrading an opponent's ability "to see, report, and process information" (Electronic Warfare 2007). This, along with other aspects of EA, can also support PSYOPs and deception. For example, control of the electromagnetic spectrum can allow transmissions and broadcasts to be hijacked and replaced with messages from the opposing side. Further, intrusion or hijacking of signals allows for false information to be planted. This deception can create confusion or fear within the target audience, or cause enemy forces to make misinformed decisions (Thomas 2009, 105). Purposefully radiating energy to create false targets, for instance, can cause an enemy to waste ammunition, be lured into a trap, or be led away from friendly forces. Lastly, electronic probing is considered a type of electronic attack. Similar to the probing in computer network exploitation, the EMS can be used to reveal the function and capability of enemy systems, such as causing anti-aircraft systems to light up (fooling them into activating) and reveal their location.

Electronic defense (ED) protects equipment, facilities, people, platforms, and sensors from enemy or friendly use of the electromagnetic spectrum which might degrade, destroy, or neutralize combat capability. The need to defend against friendly EMS usage is a result of the complex nature of the electromagnetic environment (EME) and its increasing use by military and civilians alike which can create overcrowding. Care must be given to the assignment of frequencies, electromagnetic compatibilities, and all-around spectrum management (Joint Electromagnetic Spectrum Management Operations 2012). Soldier education and training are important in minimizing an enemy's ability to use the EME against them. For example, understanding and vigilance on the battlefield can reduce unnecessary emissions which might be intercepted, jammed, or targeted by the enemy. This can include following proper equipment calibration, installation, and procedures; keeping transmissions brief and avoiding the unnecessary transmission of highly sensitive information; reducing power consumption; and using a variety of technical techniques like directional antennas and spread spectrum[17] (Electronic Warfare in Operations 2009). To prevent espionage or embedded backdoors in the production chain of EW equipment, the acquisition of EW products can be restricted to trusted vendors who have obtained certifications and met required standards. On a tactical level, ED includes the use of countermeasures which can protect against bombs and incoming missiles. Personnel and vehicles can use devices which jam the radio signals used to detonate some types of improvised explosive devices. Chaff, decoys, smokes/aerosols, and pyrotechnics/flares can be deployed to impair enemy "precision guided weapons and sensor systems" (Electronic Warfare 2007). Further, stealth technology evades electronic attack and detection by absorbing and redirecting radar waves, and by minimizing its visible and infrared signatures, such as the use of internal weapons bays and exhaust techniques. Camouflage is another type of electronic defense (Thomas 2007, 122, 296–97). Low-technology methods include hiding among civilians or using natural terrain and foliage to conceal troops, vehicles, and facilities from detection by aerial and satellite imagery (Thomas 2009, 105). Camouflage clothing, nets, and paint, which are relatively low-tech, are increasingly relying on computers to generate sophisticated design patterns and reduce thermal signatures (Thomas 2009, 104). Meanwhile, research and development (R&D) is underway on new types of camouflage such as cloaking devices and an "electric field-induced" chameleon film which can change color to suit the environment (Thomas 2009, 102; Mark 2008; Winkler 2003). Additional R&D points toward a rise in high technology and attempts to master the EME. These include the application of bionics and plasma to stealth design; adding chemical compounds to water fumes to absorb infrared, radar, and other EMS waves; the use of nano-ceramic materials to reduce radar signatures; and the use of specialized paint which can reduce a vehicle's surface temperature (Thomas 2009, 102, 107). All of these reduce an enemy's ability to sense, shoot, and command.

Electronic surveillance (ES) is the use of the electromagnetic spectrum to enhance intelligence and situational awareness. Night-vision goggles are one type of ES which is frequently portrayed in popular culture and are a technology which is becoming increasingly accessible to civilians through commercial entities. ES detects, locates, identifies, and evaluates sources of electromagnetic radiation. As examples, this can yield the location of enemy forces and provide an assessment of their equipment, or it can provide intercepted transmissions such as the previously discussed SkyGrabber software, which captured unmanned aerial vehicle video feeds. It is, therefore, a source of information which can be used to conduct electronic attack, improve electronic defense, task weapons and sensors, and make informed command decisions. The significant overlap between ES and signals intelligence (SIGINT) suggests the two terms may be consolidated in future organizational changes. In general, their current differences are their development path, or context in which they are examined, and historic areas of responsibility; ES tends to be more tactical, relies on rapid battlefield assessments, and is closely tied to EA and ED. Despite uncertainty within EW literature, it seems appropriate that the category of ES should also include all of the non-attack and non-defense aspects of utilizing the electromagnetic spectrum to enhance combat capability. In other words, ES includes the use of communication and radar systems by a military force, not only the surveillance of the opposing side's use of such systems. Like the Internet, digital communications go beyond voice to include images, text, and video. Morse code is an older, yet continuingly relevant, example of beyond voice communication, and it is noteworthy for being able to transmit through the visible portion of the EMS by Naval signal lamps on warships or submarine periscopes. A developing aspect of ES is wireless power, or the ability to transmit electronic fuel. There are two categories of this: non-radiative and radiative. Non-radiative, or near-field, can be observed in emerging technologies such as radio-frequency identification (RFID), the recharging stations for some electric vehicles, and wireless charging pads for computerized devices like mobile phones, mp3 players, and netbooks or tablets.[18] R&D into radiative, or far-field, technologies are attempting to transmit power through concentrated beams of electromagnetic energy. One proposed application is to provide refuelling for unmanned aerial vehicles.[19] If this seems far-fetched, remember that solar power is also electromagnetic radiation composed of the infrared, visible, and ultraviolet portions of the EMS.[20]

EW can be applied from land, sea, air, and space, as well as unmanned and autonomous systems. The technical details and full range of systems and platforms which use the electromagnetic spectrum in warfare are beyond the scope of this book; however, a few select examples will provide a better understanding. Moreover, examples of systems and platforms provided elsewhere in this chapter can also be applied to EW, not only the subsection to which they belong. This is due to significant overlap between terminologies; as previously stated, this chapter argues that all of these terms, including EW, are a part of NCW. Two aircraft which epitomize EW platforms are the United States' EA-6B Prowler and its replacement, the EA-18G Growler. Both of these aircraft possess anti-radiation missiles, the ability to locate enemy EMS signals, and the ability to conduct multiple forms of jamming. Further, they can not only jam signals but also intercept signals for "processing, analysis, and intelligence reporting" (Electronic Warfare Operations 2009). This means they are capable of both EA and ES. The Growler also possesses an anti-jamming capability and enhanced NCW attributes, such as improved situational awareness and improved connectivity with other sensors, shooters, and command (in theater and abroad). *Satellite Hacking: A Guide for the Perplexed* (Fritz 2013a), provides details on one large category of systems which are reliant on the EME.[21] It reveals an increasing reliance on such systems and how the global spread of commercial technology is placing them at risk. Moreover, it demonstrates how "hacking" closed military networks through the use of the EMS can appear quite different from CNO's Internet hacking.[22] The following paragraphs have been adapted from this text to provide greater insight into EW.

Satellite hacking can be broken down into four main types: Jam, Eavesdrop, Hijack, and Control. Despite some limitations, these four types of satellite hacking are significantly different and warrant precise terminology. Jamming is flooding or overpowering a signal, transmitter, or receiver, so that the legitimate transmission cannot reach its destination. In some ways this is comparable to a DDoS attack on the Internet, but using wireless radio waves in the uplink/downlink portion of a satellite network. In general jamming "requires a directed antenna, knowledge of the frequency to be affected, and enough power to override its source" (Rooker 2008). In many ways this can be considered the easiest form of satellite hacking since it can be as simple as throwing an abundance of noise at the receiver to drown out the transmission. The receiver can be the satellite receiving an uplink or a ground station or user terminal receiving a downlink. Jamming the uplink requires more skill and power than a downlink, but its range of disruption tends to be greater, blocking all possible recipients of the relay rather than a single or range-limited portion of possible downlinks (GAO Critical Infrastructure 2002). Alternatively, eavesdropping on a transmission allows a hacker to see and hear what is being transmitted. Information available online provides details on how to conduct such operations, as well as the means to purchase the appropriate equipment, and a niche subculture has emerged that is combing the skies in search of access points. [23] The third type of satellite hacking, hijacking, is the unauthorized use of a satellite for transmission, or seizing control of a signal such as a broadcast and replacing it with another for the purpose of misinformation, propaganda, PSYOPs, or terror.

Controlling refers to taking control of part or all of the tracking, telemetry, and control (TT&C) ground station, bus, and/or payload [24]—in particular, being able to maneuver a satellite in orbit. Controlling a satellite involves breaching the TT&C links. Theoretical examples include issuing commands for a satellite to use its reserve propellant to either enter a graveyard orbit or re-enter the Earth's atmosphere and burn up, or causing a satellite to rotate, so that the solar panels and antenna are pointed in the wrong directions (Norris 2010, 36; Gutteberg 1993, 12). Reaching this command and control level appears to be the most difficult of satellite hacking, since this is where security is greatest. In particular, military satellite networks often locate TT&C ground stations within military bases and they employ encryption at multiple levels. However, the military often leases commercial satellites to meet the growing needs they cannot fulfil on their own, and in these cases the satellite service provider is typically responsible for the TT&C links and satellite control ground station, with the military only securing the data links and communications ground stations (GAO Critical Infrastructure 2002). Further, weakness in the command and control of commercial satellites, such as VSAT hubs, could place military satellites at risk from collision or the creation of debris fields due to compromised control of the commercial satellites.

Jamming, Hijack, and Control are examples of Electronic Attack, while Eavesdropping belongs to the category of Electronic Surveillance. The complete document, *Satellite Hacking: A Guide for the Perplexed*, contains timelines of open source satellite hacking incidents. These include 20 cases of jamming, 12 cases of hijacking, and five instances of control. Among the types of satellites which have been jammed were broadcasting, communications, earth observation, GPS, imagery, and satellite phones. China in particular is noted for possessing a GPS jamming capability, as well as anti-jamming satellites.

ANTI-ACCESS AREA DENIAL

Anti-Access Area Denial (A2/AD) is another recurring military modernization concept which shares a strong connection to China's NCW, yet this connection has not been made clear in popular literature. A2/AD refers to the growing capability of countries to deny a foreign power "the ability to enter or operate in maritime territories adjacent to these countries" (Bitzinger 2014). A2/AD originated as part of a larger U.S. concept known as AirSea Battle, and it has since exceeded its parent concept in notoriety. Chinese officials have also begun using the term A2/AD, despite initial criticism that it portrayed China as an aggressor or that it sought to limit China's power (East Asia Security Symposium and Conference 2013). AirSea Battle (ASB) is often illustrated using a U.S.-China scenario, or to a lesser degree U.S.-Iran, however this concept is being implemented for use beyond these two regions, in part due to the proliferation of advanced weaponry and ICT which are making A2/AD a rising global phenomenon.[25] For example, high-performance aircraft and extended-range missiles with improved accuracy, effectiveness, and destructive power are no longer restricted to great powers. Not only does this threaten the application of sea power along foreign shores, it also threatens allied military ports and bases in the region. Since WWII, the United States has been able to operate largely unchallenged off the shores of its opponents. As demonstrated during the Korean War, Vietnam War, Gulf War, War in Afghanistan, Iraq War, and numerous other skirmishes, this allowed for the unhindered flow of "supplies and reinforcements" (Krepinevich 2010). It also allowed ports, bases, and aircraft carriers to serve as safe havens. These coupled with stealth technology, precision-guided weapons, and advanced sensors allowed the United States to conduct increasing portions of operations from a relatively safe distance. China is acquiring a range of powerful new military capabilities, in what some believe is an attempt to create a "no-go zone" off its shores and push the U.S. fleet back to the Second Island Chain.[26] These include anti-ship missile systems, ASATs, improved air defense networks, long-range surveillance and targeting, stealth bombers, advanced submarines, UAVs, and the overall ability to disrupt adversary C4ISR which is critical for operating at increased distances. It also includes the often cited, yet misrepresented, cyber warfare and electronic warfare. ASB addressed ways in which air and naval forces could integrate capabilities across all domains (land, sea, air, space, and cyber) to counter growing challenges to the United States, particularly anti-access and area denial (A2/AD) capabilities. ASB emphasizes "battle networks," electronic warfare, stealth, rapid response, long-range strikes, security through redundancy, and Air Force and Navy interoperability; all of which are closely tied to NCW (Krepinevich 2010; U.S. Department of Defense 2013). ASB, and A2/AD specifically, are both worth examining in this chapter, since the two influence each other and are accelerating NCW advancement on both sides.

China perceives the United States as its greatest threat, and the United States relies on NCW, therefore it seems logical that China's A2/AD would be set to counter the use of NCW. Moreover, China is developing its own NCW capabilities which it would use as a part of A2/AD. In this sense, A2/AD in a U.S.-China scenario can be seen as an NCW battle under set conditions. It would be two opponents at different NCW and military development levels, with the more advanced NCW force pushing in on the other's coast. It is ironic that China is pursuing the same NCW capability that the United States possesses, yet in this scenario that U.S. capability could provide asymmetric vulnerabilities of which China would attempt to take advantage. Even less intuitive, China would be using select NCW-enabled abilities to do so. In other words, China is not opposed to NCW, but it would attempt to exploit U.S. reliance on it and the weaknesses it contains, until China's NCW can reach a comparable level. This scenario also fits the doctrine of "local wars under conditions of informationization" (high-tech conditions), which will be discussed later in this chapter, and there are several "home field" advantages for China. Additionally, this scenario raises a larger question which is present here and beyond—will China's proclaimed defensive and regional policies continue once its NCW development (among other competencies) reaches a global capacity? Regionally China is concerned with protecting "national sovereignty, security and territorial integrity," including maritime disputes, the One-China policy, and deterring U.S. containment (Information Office of the State Council of the People's Republic of China 2013). Yet China also seeks "national economic and social development," and this requires greater interaction with the international community, the protection of SLOCs, and the ability to project power (Information Office of the State Council 2013). It is important to note that A2/AD activities are not entirely comprised of NCW. The A2/AD phenomenon is a result of the proliferation of advanced weapons and ICT. Many advanced weapons are directly related to NCW, and ICT is a driving force behind NCW, so there is a strong connection. However, employing anti-access methods does not restrict itself to NCW-only activities and can include CNO,[27] IO, and traditional weaponry used against military and civilian targets of the opposing force or its allies.

China could use A2/AD techniques to erode the ability of the United States to operate in the Western Pacific. Central to this would be precision strikes against critical nodes to render U.S. forces "deaf, dumb, and blind" (Tol, Gunzinger, Krepinevich, and Thomas 2010; US Department of Defense 2013). For example, kinetic ASATs, ASAT lasers, jamming, and space-based ASAT weapons could degrade or destroy U.S. satellites which provide a wide range of communications, early warning, navigation, surveillance, and targeting, as well as entire systems like GPS or some models of UAVs which are reliant on satellites (Chase, Engstrom, Cheung, Gunness, Harold, Puska, and Berkowitz 2015; McGarry 2013b; USCC Annual Report 2014). In the event that a conflict in the Western Pacific escalated to the point of unrestricted targeting of space assets, the PRC would enjoy the advantage of regional proximity. China could rely on its land-based, hardened, and buried fiber-optic connections, whereas the United States would be operating far from its home and be more reliant on wireless network connections. Further, China would have shorter lines of communication and a smaller area requiring coverage. However, a loss of satellites could render China's A2/AD ineffective since China needs them for over-the-horizon identification, targeting, battle damage assessment, situational awareness, and so forth. The same benefits that satellites provide the United States would aid China in its ability to conduct A2/AD, and Beijing is continually improving and becoming reliant on its own satellites. In addition to striking critical nodes, A2/AD shares another similarity with CNO—asymmetric qualities and the assassin's mace. For example, ASAT dazzlers are much cheaper than the cost of constructing, launching, and operating a satellite. One military analyst estimates the cost of one Chinese DF-21D anti-ship ballistic missile, which has the potential to "mission kill"[28] an aircraft carrier, at 11 million USD compared to the 13.5 billion USD cost of an aircraft carrier (USCC Annual Report 2014). Moreover, direct confrontation could be avoided through attacks on the logistical aspects of force deployment, disrupting an opponent's civilian society, exploiting vulnerabilities in essential networks, or coercing an enemy's allies to deny the use of their bases. Assassin's mace tactics seek to deliver a swift and devastating blow, such as an EMP or NCW-embedded backdoors, and being the first to take destructive action must be considered (Krepinevich 2010; Mulvenon 1999). Adding to non-lethal operations, unattributable actions such as mapping networks, planting latent viruses or backdoors, mapping underwater terrain, and cataloguing enemy space assets would "minimize the time required to execute attacks against key nodes and sites" at the onset of conflict (Tol, Gunzinger, Krepinevich, and Thomas 2010).

Precision strikes against crucial hubs or links can include network nodes such as aircraft carriers, airfields, bases, ports, satellites, and vital aspects of deployment and logistics. An often cited component of Chinese A2/AD is anti-ship ballistic missiles (ASBM) and anti-ship cruise missiles (ASCM), like the Russian-made SS-N-22/SUNBURN and SS-N-27B/SIZZLER supersonic ASCMs (U.S. Department of Defense 2013). Some of these missiles could be destroyed by missile defense systems, however, a missile barrage or the use of decoys could exhaust these systems' limited supply of magazines. Placing ships in adjacent waterways at risk reduces their freedom of action and slows operations; it could also force aircraft carriers to operate at such a distance that their ability to provide air support is severely degraded. This would place increased pressure on land-based runways which are even more susceptible to runway cratering as they are static targets. Anti-ship missiles can use over-the-horizon radar for targeting and utilize anti-radiation detection, microwave EMP, MirVs (decoys), and multi-spectral imaging, all of which were discussed under electronic warfare in this chapter's previous subsection. Additionally, frequently discussed components of China's A2/AD capacity include aspects of NCW, like C4ISR systems and "near real-time sensor-to-shooter" networks; and RMA aspects, like advanced submarines, AEW&Cs, long-range capabilities, missile defense, ramjet propulsion, and stealth aircraft and warships (USCC Annual Report 2007). In the naval realm, China's Kilo-class submarines possess guided weapons, such as Russian-designed "wake-homing and wire-guided torpedoes," and the ability to fire missiles while submerged (Krepinevich 2010). Undersea sensor networks can provide China with enhanced underwater situational awareness. These sensor networks could be linked to automated systems for deploying torpedos or "mobile mines"; and "acoustic jamming" could hinder an opponent's ability to operate in littoral areas or mask friendly movements (Tol, Gunzinger, Krepinevich, and Thomas 2010). In the air domain, China's Eastern coast "is blanketed by a dense network of SAM launch sites and radars," which could even threaten stealth aircraft (Krepinevich 2010). China itself is deploying stealthy fifth-generation aircraft and a number of UAVs, including the High-Altitude Long Endurance (HALE) ISR UAV. This UAV can extend the range of over-the-horizon radar, and all-around C4ISR, thereby providing an alternative to satellites for multiple A2/AD functions.

Rather than deterring U.S. development of NCW, A2/AD seems to be pushing forward and strengthening its development. For example, the uncertainty of being able to maintain unchallenged access to the Western Pacific is pushing the United States to develop unmanned submersible vehicles, directed energy missile defense systems, and the ability to apply force from even greater distances, all of which were already goals of NCW. As another example, concern that UAVs could be grounded in the event of a satellite network disruption could drive militaries toward "conducting autonomous [UAV] ISR or strike missions," since no satellite communication would be needed (Tol, Gunzinger, Krepinevich, and Thomas 2010). Many of the vulnerabilities in U.S. NCW (including RMA and EW) which China would seek to exploit in an A2/AD campaign had previously been identified by the United States itself. Adding the threat perception of Chinese A2/AD provides additional justification and budgetary support for addressing these problems. Aside from strategy adjustments, the best defense against A2/AD appears to be fixing the problems with NCW and making it as optimal as possible, rather than scrapping the concept all together. This would include creating redundant airborne C4ISR systems as an alternative or backup to satellite networks, the hardening of forward bases and computer networks, ensuring computer networks are closed and secure, enhancing system standardization and interoperability, identifying critical nodes and making those the top priority for defensive improvements, and preparing for the possibility of network outages. However, one major problem with all of the NCW categories is the cost factor, in this case A2/AD could lead the United States to overextend its resources.

In the U.S.-China scenario, the United States would attempt to conduct a blinding campaign, just as China would for anti-access and area denial. Long-range precision strikes, and the use of stealth technology and drones, would target China's C4ISR sensors (including over-the-horizon radar) and network functionality to disrupt situational awareness, communication, and targeting. Anti-ship missile, counter-space, and land-based missile sites would be high-priority targets. Some of these are mobile, so it would require timely surveillance and tracking. Using decoys, or electronically spoofing China's radar sensors, to create false targets could cause China to expend its large missile inventory and reveal valuable targets. As with China, the use of embedded backdoors and latent kill switches within NCW equipment and software could be exploited. Electronic attack and defense, such as jamming, directed energy weapons, and camouflage, could also be used to deny China's emerging UAV ISR capabilities. Acting at greater distances would challenge China's current military capability and range. In this regard, the United States could conduct "distant blockade operations" to discourage Chinese actions (Tol, Gunzinger, Krepinevich, and Thomas 2010). This is especially relevant in view of 85 percent of China's oil imports traversing through the Strait of Malacca and South China Sea, as well as the majority of its gas imports and overall trade (Le Mière 2013; Richardson 2012; Storey 2006). Additionally, undersea fiber-optic cables which electronically connect China with the rest of the world could be disabled from great distances; and the United States' extensive experience and R&D in space activities suggests they possess a premiere anti-space capability. Undoubtedly these actions would entail serious repercussions; however, they remain valid theoretical concepts for scenarios and contingency planning.

INFORMATIONIZATION

There is much confusion surrounding the term informationization. The term has multiple spelling variations, vague and varied definitions, including two distinct types of definitions among those, and overlap with multiple other terms. In some cases, Chinese and English authors are using RMA, NCW, EW, and other cyber terms interchangeably, inconsistently, or with no clear distinction (Thomas 2007, 89). Some of this stems from the PLA's use of a wide range of Western doctrinal military writings. These are selectively adopted and slightly modified, without clear reference to the original, and then converted into Mandarin (Mulvenon 1999). This creates recognizable themes and recurring elements among research material, yet there are a large number of discrepancies when comparing finer details. These problems will be explored below, ultimately arriving at the conclusion that informationization is equivalent to NCW, including RMA, EW, and the majority of A2/AD features. Once this connection is established, the previous sections of this chapter can be viewed as being as much about China as this section is. In other words, RMA, NCW, EW, and ASB are what China is developing, and China faces the same "deep-level contradictions and problems" in this regard as the United States does (Krekel, Adams, and Bakos 2012). Additionally, this section will examine China-specific literature on the development of informationization (NCW), government organization, and as yet undiscussed weapons platforms. In essence, informationization is the integration of ICT and computerized networks into "military systems and doctrine" to enhance combat capability (USCC Annual Report 2007). It is a beneficial way to "acquire, transmit, process, and use information during war" (U.S. Department of Defense 2014).

There are at least six primary variations of the word informationization, all beginning with the prefix "informa." These are: -tionalization, -tionalized, -tionization, -tionized, -tization, and -tized. Additionally, it is common to see some sources replacing the letter "z" in these words with the letter "s" ("ise" being a widely used form by British English). One reason for these discrepancies is variation in translations from Mandarin to English, and the subsequent use of these varied translated terms by Chinese authors, which are then translated from English back into Mandarin (Thomas 2004, 3). Further, it appears that leading Chinese military analysts and government officials have not settled on a term themselves. China's bi-annual national defense white papers began using informa-terms on a wide scale beginning in 2004, first as "informationalization" and then as "informationization" in 2006 and all subsequent years (see table 4.1). These seemed to replace or supplement the term modernization, and this makes sense as NCW-related advancements, or a global trend toward the use of computers in military weapons and operations, are the current direction of modernization. Despite frequent use of the words technologic, technologies, and technology in China's defense white papers, there is almost no mention of CNO, IO, the Internet, online, or the Web. This suggests that informationization is distinct from CNO and IO which are Internet-connected. NCW is not mentioned at all in these national defense white papers. Given the emphasis the United States has placed on NCW, and China's pattern of examining and integrating key U.S. military developments, this further suggests that informationization is synonymous with NCW. Meanwhile, RMA and EW are mentioned an intermediate amount of times. Beyond inferring a correlation, there are instances where PLA officers and Chinese government officials directly state that informationization is equivalent to NCW, or they refer to the United States as having an informatized military (Thomas 2007, 80; U.S. Department of Defense 2013). Unfortunately, the majority of authors are not so explicit in making this connection and instead leave it open to interpretation.

Informationization is often noted as part of a larger Chinese slogan or policy goal—the ability to win "local wars under conditions of informationization" (China's Military Strategy 2015; Information Office of the State Council of the People's Republic of China 2013; USCC Annual Report 2012). This is an updated version of the PLA's previous stated goal of being able to "fight local wars under high-tech conditions," which was in use from approximately 1993 to 2002 (Krekel, Adams, and Bakos 2012). This suggests there is possibly a strong connection between the meanings of informationization and high-tech conditions. Informationization continues to refer to modernization and high-technology conditions, but is attempting to be more precise and descriptive by conveying the central role that information plays in these, particularly through its sharing via networked systems. This is equivalent to NCW, although the term NCW is less clumsy, and carries a more intuitive definition in English. The reference to local wars is also revealing in that it corresponds to China's stated regional focus and concurrent A2/AD capability. As stated above, these policies could shift as China's NCW and overall military modern-

Table 4.1. Use of Key Terms in China's National Defense White Papers

	2000	2002	2004	2006	2008	2010	2013
Informa-terms	1	1	40	30	48	34	11
Moderniz-	20	18	14	16	21	20	7
Technolog-	43	64	81	46	50	55	10
CNO	0	0	0	0	0	0	0
IO	0	0	2	0	0	0	0
Internet	0	0	0	0	0	0	0
Online	0	0	0	0	0	0	1
Web	0	2	2	0	5	1	0
NCW	0	0	0	0	0	0	0
Network	3	6	6	7	5	8	2
RMA	0	0	10	7	8	3	0
EW	0	1	0	0	1	0	0
Electromagnetic	0	0	1	0	8	8	0

Information Office of the State Council of the People's Republic of China 2000; Information Office of the State Council of the People's Republic of China 2002; Information Office of the State Council of the People's Republic of China 2004; Information Office of the State Council of the People's Republic of China 2006; Information Office of the State Council of the People's Republic of China 2008; Information Office of the State Council of the People's Republic of China 2010; Information Office of the State Council of the People's Republic of China 2013.

ization reach a sustainable global capacity, and as China becomes entangled in worldwide responsibilities resulting from transregional needs.

An additional difficulty with clarifying informationization's meaning is that it is occasionally used by PLA officers and government organizations to mean something separate from the warfare concept described thus far, and this split meaning has to be identified by context. This chapter is discussing the more common usage which refers to "military informationization" as opposed to "national informationization" which denotes modernizing the government and society through Internet-connected information networks (Thomas 2007, 86, 88). China's Twelfth Five Year Plan 2011–2015 provides an example of the term informationization being used by Chinese officials in a broad, non-military, context. Chapter 13 is titled, "Comprehensively improve the informationization level"; however, the title is the only occurrence of the word in that chapter. The chapter content discusses non-military networks that are connected to the Internet, such as goals to enhance national broadband coverage, mobile phone networks, digital radio and television broadcasts; promote e-business and e-government services; and strengthen online IT security (China's Twelfth Five Year Plan 2011). While the word informationization is occasionally used in this context, it is the exception not the rule. One of the reasons this book prefers the term NCW is that it is less susceptible to being used in this way, in part because it directly contains the word "warfare." Chinese documents may eventually make this distinction clearer in their choice of terms, as evidenced by the recent name change of the "Internet Security and Informationization Leading Group" to the "Central Leading Group for Cyberspace Affairs" (China's New Small Leading Group 2014; Creemers 2015; Jing 2014; Panda 2014; Rawlinson 2015; Tham 2015; Xi Jinping leads Internet security group 2014). This may have been an attempt to clarify the group's function, as it deals with IO, and to a lesser degree CNO, but not NCW. As noted in the NCW section above, China possesses and is further developing closed networks for military operations (Krepinevich 2010; Thomas 2004, 10–11; Tol, Gunzinger, Krepinevich, and Thomas 2010; Wilson 2007). Adding to the NCW-Informationization connection, key features of informationization which are frequently discussed in informationization literature will be examined below to show how they align with the key features of RMA, NCW, and EW literature.

The 1991 Iraq War is frequently cited as a watershed moment, or revelation, for the RMA, and the same is true for informationization. PLA leadership was impressed by U.S. dominance, including the use of precision strikes and stealth technology, and this provided an impetus for China's push toward informationization (Mulvenon 1999; Thomas 2004, 4, 27; U.S. Department of Defense 2013; USCC Annual Report 2014; Wortzel 2014). Influential PLA authors and military strategists from China assign varying meaning to their use of the term RMA (Thomas 2007, 78).

Some directly describe it as informationization, or the use of informatized equipment, but there is no consensus. Further, these authors sometimes stray from their own definitions within the same text. Therefore, it is more useful to look for patterns rather than the details of a single definition. Recurring topics which are attached to the RMA in Chinese literature are also frequently seen in NCW, EW, and informationization literature. These include: C4ISR, communications, high technology, ICT (as a driving force), increased ability to act from a distance, increased combat efficiency and effectiveness, increased speed of actions and decisions, missile defense systems, networked systems, non-lethal weapons, precision strikes, reduced causalities, sensors, and situational awareness (Thomas 2007, 71–96; U.S. Department of Defense 2014). These authors also discuss the need for changes in organization, strategy, tactics, and theory in order to take advantage of new technologies, and they note the danger of information overload.

It is common for Chinese and Western authors to refer to RMA or NCW "with Chinese characteristics," and perhaps part of the rationale for using the term informationization is an attempt to convey something which is unique to China (Information Office of the State Council 2013). It is true that China has a different level of economic and scientific development as well as a different philosophical and cultural base from which to formulate its military concepts, rather than "simply follow Western thinking and add only Western developments and technologies to the existing framework" (Thomas 2007, 74). However, current distinction between Chinese and Western RMA-type developments is overblown and not convincingly demonstrated by prominent authors (Thomas 2004, 5–11). For example, a recurring topic which is proclaimed to be uniquely Chinese is a greater focus given to the cognitive, psychological, or perceptual aspects of warfare, yet this was already shown earlier in this chapter to be an early U.S. NCW concept, and many Western authors are discussing the "mind" aspect without explicitly labelling it as such (Ahvenainen 2003; Alberts, Garstka, and Stein 2000; Cyberspace Operations 2013; Information Operations 2012). Additionally, while some Chinese and Western authors see this as a Chinese characteristic, other Chinese writers directly give credit to the United States for its use (Thomas 2007, 90–91, 164). Technology is the driving force behind informationization, and this constrains the possibility of drastically different approaches emerging. In other words, globalization, strong ties to the commercial sector, and the universal laws of science ensure that the best approaches or network architecture will be adopted globally. Greater variation might occur in the organizational and strategy aspects of RMA-style development. Although, once this knowledge becomes known to opposing militaries, they could decide whether to adopt it themselves, and this book is based on open source knowledge. It is in China's interest to pursue its own novel ideas; however, any advantageous revelation in informationization is likely to be incorporated into NCW, and vice versa. This continual "merging of Eastern and Western military cultures" will limit country-specific characteristics (Thomas 2007, 76).

China has closely studied U.S. NCW developments and appropriated them for informationization. For example, a book published by the PLA, titled *An Interpretation of Network Centric Warfare*, uses the terms informationization and NCW interchangeably (Thomas 2007, 163–74). On the whole, the book insinuates a difference between these terms, yet it offers no clear distinction. Only the final chapter is dedicated to Chinese-based NCW, and this chapter lacks the detail of the previous U.S.-centric chapters, suggesting that the Chinese version is largely the same as that of the United States. Further to this point, "informationized weapons" have been described as those which use computer networks and "other technologies in unison," and "informatized wars" are wars which are heavily reliant on computer networks, particularly for C4ISR and situational awareness (Thomas 2007, 167; USCC Annual Report 2007). Discussion of informationization commonly features the same topics discussed in NCW, without any distinguishing differences. This includes GPS, integration, interoperability, jointness, mobility of combat forces, operating from great distances, rapid decision making, and reduced preparation time. The CMC and PLA view systems of systems, an NCW concept, as central to achieving informationization. They are seeking to link all commanders, platforms, sensors, and shooters through closed computer networks (International Institute for Strategic Studies 2015; Krekel, Adams, and Bakos 2012; USCC Annual Report 2006). Electronic warfare, a component of NCW, is also frequently discussed as part of informationization. This includes regular mention of anti-radiation missiles, electronic decoys, jamming, radar, wireless communication, and training "under complex electromagnetic environments" (USCC Annual Report 2009; Information Office of the State Council 2013; Krekel, Adams, and Bakos 2012; U.S. Department of Defense 2013; USCC Annual Report 2014).

Having established the connection between Informationization and NCW (including RMA, EW, and A2/AD), additional aspects of Chinese NCW will be examined. The General Staff Department's Department of Informationization,[29] along with individual initiatives among the military branches, are seeking to achieve integrated operations by modernizing equipment and linking systems across domains and military branches (Information Office of the State Council 2013). It is believed this will allow for leaner forces and joint operations, as well as the ability of a central command to control, direct, and coordinate a greater portion of operations by using "information networks to transmit battle space awareness data and joint strike commands" (USCC Annual Report 2007). In China's pursuit of a system of systems, it is using a combination of "fiberoptic cable, high-frequency and very-high-frequency communications, microwave systems, and multiple satellites" to facilitate connections (Wortzel 2014). The U.S. Department of Defense 2013 annual report to Congress on China's military capability provides insight into China's current level of regional informationization capability.

> New technologies allow the PLA to share intelligence, battlefield information, logistics information, weather reports, etc., instantaneously (over robust and redundant communications networks), resulting in improved situational awareness for commanders. In particular, by enabling the sharing of near-real-time ISR data with commanders in the field, decision-making processes are facilitated, shortening command timelines and making operations more efficient. These improvements have greatly enhanced the PLA's flexibility and responsiveness. "Informatized" operations no longer require meetings for command decision-making or labor-intensive processes for execution. Commanders can now issue orders to multiple units at the same time while on the move, and units can rapidly adjust their actions through the use of digital databases and command automation tools. (U.S. Department of Defense 2013)

China possesses an Integrated Air Defense System (IADS) and early warning sensors capable of regional coverage. These consist of "weapon systems, radars, and C4ISR platforms working together" with a primary goal of defending against long-range precision weapons and manned and unmanned stealth aircraft (U.S. Department of Defense 2014). The C4ISR component relies on "reconnaissance, data relay, navigation, and communications satellites," although the aircraft discussed below could provide alternatives (U.S. Department of Defense 2014). China's strategy is to concentrate its "best capabilities against the enemy's most important assets" (Krekel, Adams, and Bakos 2012). Moreover, using long range strikes against critical nodes could deter an adversary's willingness to fight, without the need for mass destruction or "the seizing of territory" (Andress and Winterfeld 2011, 43; Mulvenon 1999).

In addition to keeping these networks separate from the Internet, the tangible infrastructure must be protected. For example, fiber-optic cables can be buried, and facilities can be camouflaged, hardened from blasts, and shielded from electromagnetic radiation. To this end, China's drive toward informationization has accelerated the construction of underground facilities and research into "advanced tunnelling and construction methods" (U.S. Department of Defense 2013). Weapons platforms are often required to remain mobile, however some command and control facilities might remain static, and China's no-first-use policy necessitates that they are able to survive an initial nuclear or EMP blast. As with A2/AD, China would not only use informationization to enhance its abilities, it would seek to deny an opponent's ability to do the same. Another consideration is that all of the problems associated with developing NCW-style capabilities, which were listed in the previous sections of this chapter, apply to China. The technologic, organizational, and financial challenges of integrating such a wide range of systems, and relying on these to provide reliable and timely information, has the potential to backfire.

China's rapid economic growth, promotion of science and technology education, and attempts to transition to high-technology exports benefit informationization. China's military budget has increased by double digit percentages every year since 1989, with the exception of 2010 (USCC Annual Report 2014). In 2014, China's military budget was approximately 131.6 billion USD, which is second only to the United States at approximately 581.0 billion USD (International Institute for Strategic Studies 2015; USCC Annual Report 2014). However, China's total military spending may be far greater than the official figures reported. Foreign acquisitions, research and development of dual-use science and technology, national security, construction, and emergency response and disaster relief are a few examples of expenditures which may fall under non-military headings but directly relate to the advancement of the military (Fritz 2008). Additionally, China may be saving billions of dollars in R&D and leapfrogging in development due to alleged technology transfer through computer network exploitation, traditional espionage, and reverse engineering. The State Council's Fifteen Year Plan for science and technology (2006–2020), states that China's "defense industries are pursuing advanced manufacturing, information technology, and defense technologies," including "radar, counter-space capabilities, secure C4ISR, smart materials, and low-observable technologies" (U.S. Department of Defense 2013). In addition to the PLA's clear push towards R&D and training in "new and high technology weaponry," national-style informationization development aids military-style informationization (Information Office of the State Council 2013). While China's Twelfth Five Year Plan has a different use for the term informationization than how it is discussed in this chapter, dual-use military application needs to be considered. ICT professionals can be recruited into the military, as can their research and products which would have the added ease and security of being domestically produced. China's push towards civilian innovation in science and advanced technology includes key areas such as "new-generation IT, biology, high-end equipment manufacturing, new energy sources, new materials," and new transportation systems (China's Twelfth Five Year Plan 2011). Further, Chinese state-owned enterprises (SOEs) allow the CCP to give preferential treatment to the development of key defense-related industries, in an attempt to create "national champions." Industries that the Chinese government has identified as "strategic" and "heavyweight" include armaments, automobiles, civil aviation, construction, fossil fuels, machinery, metals, power generation and distribution, shipping, and telecommunications (USCC Annual Report 2011).

PLA modernization has focused on quality over quantity with many new platforms possessing over-the-horizon weapons and sensors (USCC

Annual Report 2014). As has previously been stated, modern military weapons often utilize computers and computer networks, therefore this book cannot cover all of them in detail. However, three types of Chinese aircraft warrant further examination, since they are frequently mentioned in association with informationization literature. The first is China's fifth-generation stealth fighters, the Chengdu J-20 and Shenyang J-31. Once fully operational, these aircraft are expected to possess "advanced sensors, radars, and datalinks" as well as electronic attack and defense capabilities (USCC Annual Report 2014; U.S. Department of Defense 2013). Second is China's Airborne Early Warning and Control (AEW&C) aircraft. These can extend the coverage and range of ground- and space-based C4ISR platforms, or provide an alternative to them. China's primary AEW&Cs are the KJ-2000 and the smaller KJ-200 (or Y-8) (U.S. Department of Defense 2014; USCC Annual Report 2014). Lastly, China has been fielding a number of unmanned aerial vehicles and unmanned combat aerial vehicles dating back to at least the 1990s with the Israeli-made IAI Harpy UCAV (U.S. Department of Defense 2013). Five prominent Chinese designs are the Harbin BZK-005, Lijian (or AVIC 601-S), Sky Saber, Xianglong (or Guizhou) Soaring Dragon, and Yilong (or CAIG Wing Loong) Pterodactyl, with the Lijin being a candidate for future carrier-based launches (U.S. Department of Defense 2013; USCC Annual Report 2014). Collectively these UAVs possess qualities of extended range, long endurance, and stealth. They can enhance multiple aspects of C4ISR, including extending the range of over-the-horizon radar, and providing surveillance and battle damage assessment (Tol, Gunzinger, Krepinevich, and Thomas 2010). Further, they can allow China to conduct long-range strike operations, aid in anti-ship missile targeting, and conduct electronic attacks. Given the relatively low cost of these UAVs, they could provide an appealing means for managing China's maritime disputes (USCC Annual Report 2014).[30]

As China's satellite capabilities expand, so does its dependence on them. Initially China sought "fiber optic cables, mobile radios, datalinks, and microwave systems" for its C4ISR, however, over the decade 2004 to 2014 this has shifted towards space-based assets (USCC Annual Report 2014). The PLA views control of space systems "as central to enabling modern informationized warfare" (U.S. Department of Defense 2014). Space is its own domain of warfare, however, the non-Internet-connected military satellites are largely tied to the cyber domain's NCW. In essence, these satellites are all nodes in various computer networks, because they must transmit the information they receive or collect back to Earth. As the space domain develops it will expand to take on a greater variety of missions beyond satellite data relays. The land, sea, and air domains are also becoming increasingly entangled with NCW. This, however, represents a more gradual process compared to the challenges of operating in space that require it to be a high-technology computerized endeavor from its inception. As of 2014, China has "approximately 100 active satellites in orbit" with many performing C4ISR functions (USCC Annual Report 2014). This includes the ability to provide communications, data relay, electronic surveillance, imagery, long-distance targeting, meteorological information, navigation, radar, and situational awareness. Nonmilitary communications and scientific satellites can be appropriate for military purposes or serve dual-use functions, as can manned spaceflight missions. China is developing a range of counter-space capabilities including direct-ascent ASATs, directed energy weapons, and jammers; as well as co-orbital ASATs capable of colliding, damaging, docking, grabbing, jamming, or mine laying (USCC Annual Report 2014). The majority of China's maritime ISR satellites are currently focused on the First Island Chain, however this is expected to expand to the Second Island Chain and Indian Ocean, and eventually provide global coverage. Similarly, China's Beidou satellite navigation (and communication) system is planned to expand beyond its present regional coverage.[31]

The PLA regularly conducts training exercises under "conditions of informationization," including elements such as C4ISR, jamming, joint operations, and precision strikes (Information Office of the State Council 2013). There are numerous PLA, Chinese government, and Chinese media reports discussing information-related training exercises. However, the language used in these reports is often imprecise, making it difficult to determine whether these were NCW or CNO exercises, and in many cases they were both. For example, a 2008 PLA training exercise employed "simulated cyber [CNO] and electronic attacks [NCW]," and since 2008, all large-scale PLA exercises have included "cyber [CNO] and information-operations [NCW] components" (USCC Annual Report 2009; International Institute for Strategic Studies 2015). The context and key words reveal which branch, or branches, are being tested. It appears that training "under complex electromagnetic environments" became a core requirement[32] for PLA training around 2007, with the addition of "system of systems operations" around 2010, both of which belong to the informationization (NCW) branch of cyber warfare (USCC Annual Report 2009; Krekel, Adams, and Bakos 2012). Military exercises conducted under informatized conditions occur "nearly every week or month by units somewhere in China" (Thomas 2009, 104). Recurring activities observed in these exercises include the use of C4ISR, electronic decoys, EMP attacks, "jamming communications and radar systems," interoperability, stealth, and attempts to strike vital points early in a conflict in order to obtain information dominance (USCC Annual Report 2009). These exercises demonstrate China's informationization abilities and goals, as well as the problems they face. For example, Chen Weizhan, the head of the Guangzhou MR's "Military Training and Service Arms Department," stated that extensive software incompatibilities, "unmatched hardware interfaces," and "non-unified data formats" significantly hindered combat operations and joint exercises (Krekel, Adams, and Bakos 2012).[33]

CONCLUSION

By examining all five primary and interwoven concepts of networked warfare, a more complete understanding of China's NCW capability and development is provided. The Revolution in Military Affairs (RMA) is a theory which claims future warfare will witness a dramatic increase in combat effectiveness and efficiency through the application of presently developing technologies. Net-Centric Warfare (NCW) is the application of ICT and advanced technology to link military (command, sensor, and shooter) systems together in order to achieve a battlefield and information advantage. Electronic Warfare (EW) is a sub-component of NCW which aims to enhance combat capability through dominance of the electromagnetic spectrum. Computerized communications, radar, and wireless closed network connections are among the uses of the electromagnetic spectrum. NCW and its accompanying technologies are the primary means through which China could conduct Anti-Access Area Denial (A2/AD). A2/AD would attempt to deny an unauthorized foreign military access to areas within the First Island and Second Island Chains, and this fits China's stated goal of being able to win local wars under informatized conditions. Informationization is China's present modernization goal, and it seeks to increase combat capability through the use of battle networks and high technology. All of these concepts involve the use of closed military computer networks which have the potential to be hacked.

Chinese government and military publications tend to use the term informationization; however, this book prefers the term NCW for multiple reasons. First, the many varied definitions of RMA, NCW, and informationization are largely synonymous, and EW and A2/AD are subcomponents of them. Prominent Chinese and English publications use these terms interchangeably, inconsistently, or with no clear distinction. Second, informationization itself has vague and varied definitions with at least six major spelling variations—there is no consensus. It is also used to represent two distinct concepts, military informationization or national informationization, which must be determined by assessing the context. Further, there is some evidence to suggest this name will be changed in the relatively near future. Third, informationization and NCW in particular appear to be interchangeable. There are instances where Chinese documents label the U.S. military as being informatized, define informationization with the same definitions as NCW, or explicitly state that informationization is the same as NCW. Part of the confusion could be the result of the PLA integrating concepts from Western military writings.

Other writings which proclaim the existence of a unique NCW "with Chinese characteristics" fail to account for the impact of globalization and the scientific method. Fourth, this book has chosen to use the term NCW as it has a more intuitive meaning in English, and it predates the term informationization.

NCW-related literature often contains small collections of lists, or recurring key terms, which represent examples and core attributes of interlinked systems. They can be of assistance in capturing the essence of NCW. Among these are broad types of existing and overlapping systems which could be incorporated into a larger system of systems. These include anti-ship missile systems, autonomous or semi-autonomous systems, C4ISR systems, early warning systems, global positioning systems, integrated air defense systems, missile defense systems, and satellite systems. Many of these enhance situational awareness and improve decision making through a greater availability of information. Specific types of weapons and weapons platforms are also commonly identified, like anti-radiation missiles, ASATs, directed energy weapons (EMP, laser, or microwave), non-lethal weapons (aiding a reduction in casualties and collateral damage), precision-strike weapons, and stealth technology (with low infrared, radar, and visible signatures). As discussed in this chapter, U.S. weapons platforms which are representative of NCW were the EA-6B Prowler, EA-18G Growler, Global Hawk UAV, Predator UCAV, Reaper UCAV, X47-B carrier-based drone, and upgraded versions of the Bradley Fighting Vehicle and M1 Abrams tank. Three measurable attributes or benefits of NCW systems are distance, size, and speed. NCW seeks the ability to operate from great distances, such as extended-range missiles with improved accuracy, effectiveness, and destructive power. Additionally, over-the-horizon radar, long-range surveillance and targeting, and geographically dispersed collaboration or command and control all have this distance factor. Size reductions of combat units and logistical requirements due to the implementation of NCW allows forces to travel lighter and faster, and it increases mobility and the ability to deploy quickly. Speed in itself is a key attribute, with NCW enabling increased tempo and lethality of operations, speed of command, speed of communication, rapid target assessment, rapid decision making (or response), and reduced preparation time. Lastly, NCW literature notes recurring themes of coordination, integration, interoperability, and joint operations.

All of these new technologies and systems provide new targets and methods for attack. This, coupled with the global proliferation of new technologies, and the ability to conduct operations from long distances, has expanded the battlefield. Vital nodes within networks are high-value targets, and blinding campaigns aimed at eliminating sensors can erode an NCW force's ability to operate. NCW is not only a collection of networked technologic advances, but also changes in doctrine, training, and organization which can allow the optimal use of those new technologies. Additionally, networked systems not only enhance direct fighting but also the planning, training, and logistical aspects of military operations. While there are many benefits to pursuing NCW, there are also concerns. One of these is the initial, and possibly continued, high financial cost. High-technology systems are expensive and rapidly changing. Further, force-wide restructuring is slow and it is uncertain which technology will endure as being revolutionary. Other problems such as allied interoperability, increased bandwidth demands, latency, and system security require substantial funding to resolve. The complex and interconnected nature of NCW systems could provide new and asymmetric ways for adversaries to attack. New means of attack include embedded backdoors, introducing false data, and targeting the computerized logistical aspects of force deployment. Asymmetric means of attack include the financial cost of ASATs or lasers versus satellites, anti-ship ballistic missiles versus aircraft carriers, or an electromagnetic pulse versus an NCW-reliant force. Over-reliance on NCW is a risk in itself as technical difficulties or interruptions to service could halt operations. Another concern is that overly complex systems or interfaces could cause information overload or slow productivity. Finally, autonomous systems, drones, non-lethal weapons, and precision strikes are creating new international standards and ethical concerns for warfare.

As this chapter has shown, NCW is a component of cyber warfare, because it involves computers and hacking; however, it remains a distinct branch. Using microwaves or radio waves to access the computers used by aircraft, satellites, ships, or tanks is quite different than what is commonly thought of as hacking. Further, defensive aspects such as physical camouflage seem far removed from computer networks, yet it can alter the data collection of enemy radar and satellite imagery which are components of computer networks. In some cases, the use of fiber-optic cables or removable data storage allow for non-wireless or non-proximity restricted hacking which is more reminiscent of CNO, yet these remain non-Internet networks. Another distinguishing feature is that unlike Internet-connected CNO, where computers can serve a wide range of functions and are not seen as a military threat until used for attack or exploitation, the closed networks of NCW are explicitly tied to military operations. Additionally, physical destruction under CNO has thus far been minimal and largely restricted to making other computers and networks inoperable. There are theoretical examples of derailing trains or overloading generators to cause explosions, but these have a much lower probability of occurring. On the other hand, NCW computer networks have weapons platforms as part of their network and are capable of causing mass casualties, albeit far less discrete. The wide range of military systems, each with highly technical and continually changing specifications, in addition to the classified nature of such systems, make it difficult to provide a comprehensive examination of NCW hacking.

China possesses a regional, but rapidly expanding, NCW capability. China's military buildup and modernization efforts are closely tied to NCW, as NCW-type systems are the standard for modern equipment and China has pursued quality over quantity since the 1980s. As such, China's largely regional air and naval power projection correlate to NCW capability. These also fit China's present defensive, sovereignty, and non-expansionist policies, and it fits the stated goal of being able to win local wars under conditions of informationization. This is in part why examination of A2/AD is important, because A2/AD fits China's present NCW capability and concerns. China possesses a large number of military computer systems which will become a part of their system of systems. Large-scale examples of these existing systems include C4ISR and integrated air defense. China has been developing closed networks specifically for classified information sharing, command and control, navigation, precision targeting, and tactical data links. Key Chinese NCW weapons include anti-radiation missiles, anti-ship missiles, ASATs, directed energy weapons, EMP, and jammers, while key platforms include advanced submarines, the J-20 and J-31 stealth fighters, KJ-2000 and KJ-200 AEW&Cs, and a number of UCAVs. China currently operates around 100 satellites, which provide communications, data relay, electronic surveillance, imagery, long-distance targeting, meteorological information, navigation, radar, and situational awareness. The majority of these assets are aimed at functioning out to the First or Second Island Chains. Exercises and training under NCW conditions became a core priority of the PLA around 2007, with less-focused efforts dating back to the 1990s. Key strategies which suit China's current NCW capability include asymmetric tactics and targeting critical nodes. China's increasing defense budget, promotion of science and technology education, and promotion of high-technology industries will further aid the development of NCW.

NOTES

1. One difficulty with viewing the electromagnetic spectrum and EW as their own domain is that there is significant overlap with the existing domains. Air, land, sea, and space all utilize the electromagnetic spectrum for communications, operations, or weaponry. Linking EW to its parent category of NCW and the even larger category of cyber warfare reduces the risk of creating an inefficient overabundance of domains. Further, cyberspace and the electromagnetic spectrum share the core similarities of utilizing computer networks, having an information foundation, being susceptible to hacking, and being an intangible dimension (U.S. Department of Defense 2014).

2. Computer network attack, defense, and exploitation all require the same equipment and skillset, whereas electronic warfare requires different equipment and training. Further, computer network exploitation can transition seamlessly to attack. Therefore, it seems unproductive to separate the attack portion and place it under another department. INEW might yield some novel combinations, yet these can still occur

without direct departmental grouping, and those combinations are not necessarily any better than other combinations which might occur from the random pairing of branch subcomponents.

3. If RMA is viewed as a recurring phenomenon, then there are multiple RMAs throughout history. This section is focused on examining the current RMA, but will refrain from continually labelling it as "the current" RMA throughout.

4. The Prussian military theorist Carl von Clausewitz viewed centers of gravity as focal points, the disruption of which can cause a system to collapse (Clausewitz 2013; Echevarria 2002).

5. BitTorrent is a peer-to-peer file-sharing protocol. Rather than download a file from a single source, BitTorrent allows a user to download pieces of a file from multiple sources simultaneously, in a swarming technique, and assemble those pieces into the complete file.

6. Being computerized is a key feature of NCW, otherwise "information networks" have a much longer history. A ground unit using radio to call in close air support could be viewed as an information network. Similarly, satellite and aerial imagery has been used in battle planning since their inception, the difference in NCW is the speed with which this information can be obtained, digested, shared, and implemented into planning due to ICT advancements.

7. A laser can be pointed at a target by ground forces. Known as "painting a target," the laser radiation scatters off from the target and provides a source for the missile to lock on to.

8. The aforementioned cloud computing (see chapter 1) would suit this purpose and provide redundancy, if a secure non-Internet version is developed.

9. Tactical data links can provide "beyond voice" information sharing, including command assignments, imagery, radar and target information; the operator's health status and vital signs; and the platform's status, such as real-time armament inventory, diagnostic, and fuel-consumption information.

10. Examples of varying networking programs across U.S. service branches include Advanced Tactical Targeting Technology, Air Force Link 16, Army Force XXI Battle Command Brigade and Below, Army Warfighter Information Network and Joint Network Nodes, Joint Tactical Radio System, Navy Cooperative Engagement Capability, Standardized Tactical Entry Point, Transformational Communications Study, and Transformational Satellite Communications (NCW Roadmap 2007; Raduege 2004; Wilson 2007).

11. Those familiar with American football can relate this to the two-point conversion playbook.

12. These may or may not be Internet connected. They seem appropriate to discuss under NCW, however, as they are closely tied to the battlefield. Further, where Internet connectivity does exist, a transition to closed networks is likely being perused, because adversary eavesdropping on these communications could compromise mission success (Wortzel 2014).

13. The electromagnetic spectrum includes items which are technically relevant to CNO, since it includes Wi-Fi and all that is visible to the human eye; however, authors discussing electronic warfare are overwhelmingly referring to the NCW aspects of cyber warfare.

14. The United States, European Union, and NATO have assigned letter designations to widely used frequency ranges within the radio spectrum. Some generalizations can be made as certain applications tend to group within specific bands, such as roving telephone services and broadband communication using the C and Ka-bands, respectively. "Ultra-High Frequency, X-, and K-bands have traditionally been reserved in the United States for the military" (Space Security Index 2012). Higher frequencies (shorter wavelengths) are capable of transmitting more information than lower frequencies (longer wavelengths), but require more power to travel longer distances.

15. There is some use of fiber-optic cables in NCW, but they are restricted to land and sea. Cables by themselves cannot be quickly deployed, connect mobile platforms, or offer a comprehensive ability to act from great distances.

16. China is alleged to have used a dazzler in 2006 to blind temporarily a U.S. satellite. Further, China allegedly possesses "free-electron and chemical oxygen-iodine high energy lasers" which could blind satellites temporarily or permanently (USCC Annual Report 2007).

17. Spread spectrum is a method that intentionally varies a transmission's frequency to reduce an opponent's ability to intercept or jam the signal.

18. RFID ties in with the logistics component of NCW discussed above, such as logging soldier and supply manifests or ordering and tracking supplies. Charging pads could also have increased relevance as more soldiers begin using tablet-like devices on the battlefield.

19. This technology could be combined with the ability for ships to "pass" missiles to UAVs (mentioned above under *The Development of Net-Centric Warfare*). This would significantly improve the amount of time a UAV can loiter over potential targets. Energy can be beamed to the UAV as needed, and once a target is acquired, a ship can loft a missile into the war zone, where it is "picked up" by the UAV and more precisely guided to its target. Eventually the ship itself could become an unmanned system, allowing a greater portion of soldiers and commanders to operate from greater distances.

20. Wireless power provides another example of overlap between terminologies. Powered exoskeletons (mentioned under the subsection *RMA*) utilize on-board computers and could aid in logistic support as well as combat (discussed under *The Developments of NCW*). One of the primary difficulties in developing exoskeletons has been their power source, hence the tether commonly seen in demonstrations. Charging stations or radiative wireless power (discussed here under *EW*) could provide a solution.

21. The predominant means of satellite transmission is radio and microwave signals. Any platform which resides in the air or space domains will require the EMS to communicate as part of a terrestrial network. For the land and sea domains, it depends on the type of platform and desired outcome. These domains at least have the option of wired or tangible connections, like fiber-optic cable, to create networks, whereas air and space presently do not.

22. Portions of the Internet rely on satellites and EMS connections, so CNO also exists in this realm. However, it is less prevalent in CNO than NCW, and less acknowledged in CNO literature. The intended target, and whether or not there was Internet connectivity within the network, remain decisive factors in determining if an incident belongs to CNO or NCW.

23. NCW operations do not involve the Internet, but the Internet can provide information and tools which will aid in the ability to conduct such operations.

24. The payload is usually a collection of electronic devices specific to that satellite's desired function. For example, a surveillance satellite would contain imaging equipment, while the payload for a communications satellite would include transponders for receiving and relaying signals such as telephone or television. The bus is the platform housing the payload; this includes equipment for maneuvering, power, thermal regulation, and command and control (Wong and Fergusson 2010, 35–36).

25. The spread of technology also played a role in the previous sections on RMA, NCW, and EW.

26. Chinese military theorists have identified two island chains as holding key strategic value. The First Island Chain runs approximately from "the Japanese main islands through the Ryukyus, Taiwan, the Philippines, and Borneo," while the Second Island Chain "stretches from the north at the Bonin Islands southward through the Marianas, Guam, and the Caroline Islands" (Tol, Gunzinger, Krepinevich, and Thomas 2010).

27. CNO is a potential component of A2/AD; however, A2/AD literature which discusses cyber attacks is often erroneously implying CNO when they are actually discussing NCW, because most of the networks they refer to would not be Internet-connected.

28. An anti-ship ballistic missile would not have to completely destroy its target to have the desired effect. Inflicting sufficient damage could render an aircraft carrier unable to perform its mission function or cause it to retreat.

29. This department is also sometimes referred to as the Headquarters of the General Staff (GSH) Informationization Department.

30. Monitoring all air and sea traffic in the Western Pacific alone is a "formidable challenge" (USCC Annual Report 2014). It would require near real-time identification and tracking of all vessels in a heavily trafficked 875,000- to 1.5-million-square-nautical-mile area.

31. For more detail on China's space-based capabilities, see Chase, Engstrom, Cheung, Gunness, Harold, Puska, and Berkowitz 2015; Fritz 2013a; U.S. Department of Defense 2014; and USCC Annual Report 2014.

32. Training of this type occurred much earlier, but this is when it gained prominence. Additionally, CNO continues to also be a focal point of training exercises, but this chapter is specifically examining NCW.

33. For a detailed assessment of Informationization exercises, training, and war games, from 1997 to present, see Information Office of the State Council 2013; Krekel, Adams, and Bakos 2012; Thomas 2004, 24–27, 65–72; and USCC Annual Report 2009.

FIVE

Application of Cyber Warfare to the Taiwan Issue

Applying Chinese cyber warfare to a particular issue will demonstrate how CNO, IO, and NCW work in unison, yet remain distinct. China has several maritime disputes, from Japan and South Korea in the East China Sea and Yellow Sea, to the Philippines and Vietnam in the South China Sea. The status of the Republic of China (Taiwan), however, warrants particular attention as it is a perennial affair. It has been described as "the world's most dangerous flashpoint" and the most likely to cause a Sino-U.S. conflict (Pillsbury 2004; China's Proliferation Practices 2008; Recent Developments in China's Relations with Taiwan and North Korea 2014). China views the Taiwan situation as a core national interest, and it seeks peaceful reunification under the "one country, two systems" principle (China's Military Strategy 2015). At the same time, Chinese policy dictates that it "will absolutely not allow Taiwan to be separated from China and will definitely make no promise to give up the use of force" (U.S. Department of Defense 2012). As a demonstration of this resolve, the PLA has positioned approximately 1,200 short-range ballistic missiles (SRBM) along the Taiwan Strait. A few hundred kilometers are all that separate these missiles from Taiwan's 23 million inhabitants. The United States, on the other hand, remains committed to the 1979 Taiwan Relations Act, which states that the United States must intervene militarily if China attacks or invades Taiwan, and the United States must maintain "Taiwan's self defense capability" (Pillsbury 2004). The Taiwan situation is further complicated by U.S. adherence to the One-China Policy. This means Washington must tread carefully in all decisions pertaining to the matter, so as not to offend either state.

Officially, Beijing, Taipei, and Washington are all opposed to "any destabilizing unilateral changes to the status quo," yet there is continual jostling (U.S. Department of Defense 2014). Each change in government administration among these three brings renewed uncertainty. Further, there is a high risk for incidents to escalate, and under heightened circumstances, there is minimal reaction time. As Mark Stokes of the Project 2049 Institute stated in 2010, "every citizen of Taiwan lives within seven minutes of destruction" (Taiwan-China: Recent Economic, Political and Military Developments across the Strait, and Implications for the United States 2010; Stokes 2010). In the event of a conflict, or preferably, to prevent a conflict, China will seek to leverage its cyber capabilities. On a strategic level, China maintains its pursuit of national sovereignty, territorial integrity, and peaceful development, including active defense and the development of informationization (China's Military Strategy 2015; Information Office of the State Council 2013). In relation to the Taiwan issue, the two most likely operational parameters are provided below to show how cyber warfare would be implemented, including an overview of tactical options. Lastly, cyber warfare is not restricted to China, so cyber defense considerations, which have not been explored thus far in this book, will be examined.

VICTORY WITHOUT WAR

China will not accept Taiwanese independence, and it has openly shown its capability and willingness to use force in this regard; however, Beijing does prefer a peaceful solution. This is not only a reflection of China's proclaimed culture, including harmony and defense, or even Sun Zi's deception, it is a logical preference. A full-scale missile bombardment and amphibious invasion "would strain China's untested armed forces," and it is the type of operation which would elicit "the most serious U.S. military intervention" (Taiwan-China: Recent Economic 2010; U.S. Department of Defense 2012). If China were to fail in its attempt, Taiwan would achieve independence. If China were to prevail, the death and destruction would result in global economic damage, international condemnation, and a volatile Taiwanese population. Cost-benefit analysis may change over time, and maintaining a brute-force option provides a deterrent; however, current circumstances warrant a more subtle approach. Cyber warfare can accommodate for this. For example, by using information operations, China can attempt to shape Taiwanese public opinion in favor of reunification. At the same time, computer network exploitation can be used to create circumstances which are more favorable to China in the event of a conflict. The founding father of Chinese IW, Shen Weiguang, explains these aspects of cyber warfare by citing former U.S. President Richard Nixon. In Nixon's book, *The Real War*, he states "there is a gray area between peace and war, and the struggle will be largely decided in that area" (Thomas 2004, 36).

As of 2014, the most recent polls revealed 80 percent of Taiwanese were opposed to reunification, 60 percent identified themselves as "Taiwanese" only (in contrast to "Taiwanese and Chinese"), and 50 percent viewed China's attitude toward Taiwan as "unfriendly" (Hammond 2014; Southerland 2014b; U.S. Department of Defense 2012). Additionally, a 2013 survey found that the majority of Taiwanese favored the "status quo in cross-Strait political relations for the time being" (Southerland 2014a). China can attempt to change these opinions through the use of online propaganda and soft power. Marketing can be conducted to boost China's image, highlight the cultural and economic benefits of closer relations, and portray favorable and transparent government policies. Increases in travel and educational exchanges, along with providing special treatment to Taiwanese celebrities and influential individuals, can translate into Taiwanese people voluntarily, and unknowingly, conducting IO on China's behalf. Anti-U.S. commentary and news can also be promoted. Commentators, like the 50 Cent Party, can be used to post comments which are favorable to China, and automated scripts can be used to artificially inflate "view counts" and the "thumbs up, thumbs down" voting of social media. Shen Weiguang's 2003 book, titled *Deciphering Information Security*, proposed university courses for Military Information Security Studies. These courses included: computer virus program design and application, a study of hacker methods, information attack and defense tactics, and an introduction to U.S. and Taiwan Social Information Systems (Thomas 2004, 154–55). The last course named demonstrates the importance placed on IO and China's view that it belongs to the same field as CNO.

During election cycles, China can seek to influence voters through the purchase of online advertisements and by encouraging (through electronic means) Taiwanese business workers in China to vote for Beijing's preferred candidates (Southerland 2014b). Despite an increase in skepticism toward China, public concerns over military action against Taiwan have decreased (Murray 2013; Pillsbury 2004). China will seek to continue, or enhance, this perception as it makes it difficult for Taiwan to justify its defense budget, which already has decreased from 3.8 percent of gross domestic product in 1994 to 2 percent in 2014 (USCC Annual Report 2014). Further, Taiwan has gradually been converting to an all-volunteer armed force, which is facing difficulties in its ability to recruit high-caliber volunteers and a cultural bias against military service (Murray 2013). By promoting online news articles which portray Taiwan's military, or a career in the armed forces, negatively, China can exacerbate the situation. China can also conduct information operations against the United States in order to impact Taiwan. China adamantly voices its opposition of U.S. arms sales to Taiwan through traditional and online means, with the two augmenting each other, and this has had an effect. Washington has been accused domestically of becoming overly cautious in its dealings with

Taiwan out of concern that it will offend China and disrupt economic stability (Hammond 2014; Taiwan-China: Recent Economic 2010). Similarly, this type of control, and the One-China Policy, limits U.S. military official's interaction within Taiwan. In the event of a conflict, commanders could suffer from limited firsthand familiarity of the surroundings, directly impacting on the NCW component of cyber warfare (USCC Annual Report 2014). Lastly, IO can be used to recruit Beijing sympathizers, either to enhance operations, such as computer network exploitation or technology transfer of NCW equipment, or to embed in Taiwan agents who are capable of conducting PSYOPS or sabotage in a time of conflict. The latter is connected to cyber warfare in that agents are recruited through IO.

One way to entice voluntary reunification is through increased dependence. Under Taiwanese President Ma Ying-jeou, commercial, cultural, and educational ties with China have flourished (Southerland 2014a). Taiwan has struggled to recover from the global economic crisis of 2008–2009, and this facilitated agreements with China, including the Economic Cooperation Framework Agreement signed in 2010 (Hammond 2014; USCC Annual Report 2013).[1] Not only do these agreements help in Taiwan's integration, they also provide China with opportunities for enhanced cyber warfare tactics. The 2014 Sunflower Student Movement, which protested the Cross-Strait Service Trade Agreement (CSSTA), reveals how economic arrangements can provide China with increased access and influence (Hammond 2014; USCC Annual Report 2014). Protesters argued that the trade agreement opened up Taiwanese telecommunications, publishing, and retail (through the favoring of big businesses) to increased Chinese control, thereby threatening Taiwan's national security, freedom of speech, and economy. Advocates of the agreement argued that unspecified details, which would be favorable to Taiwan, would be worked out after the treaty was ratified. The CSSTA might represent a recurring tactic used by China in which it takes "two steps forward, one step back."[2] Additionally, China has pressured media owners to censor content, and encouraged the purchase of media outlets by pro-China business people by rewarding or withholding the prize of economic opportunities on the mainland (USCC Annual Report 2014). Botnets or nationalistic hacker groups can be used to conduct DDoS attacks as a means of punishing China's critics. This provides plausible deniability for Beijing, and it encourages self-censorship. Taiwan has sustained persistent attacks from Chinese hackers dating back to the large-scale cyber conflict of 1999 (see chapter 2). PLA exercises, such as one in 1997 which the Taiwan Central News Agency labelled as an attempt to "develop a computer-virus warfare capability," can be viewed as a cyber deterrent or IO intimidation (Thomas 2004, 24). *The Science of Military Strategy* advocates using strategic deterrence to control the Taiwan situation (Thomas 2004, 305).

During peacetime, computer network exploitation can be used to alter the conditions of a potential conflict in China's favor. This includes the theft of intellectual property and defense information, as well as implanting backdoors in networks and microchips, both of which can enhance CNO and NCW. Relevant to security concerns of embedded backdoors, three of Taiwan's top five imports from China are computers, microchips, and mobile phones. Conversely, and possibly a concern for reverse engineering, three of Taiwan's top five exports to China are microchips, printed circuit boards, and semiconductors, which are tested and assembled into products in China (USCC Annual Report 2014). Since 2011, concerns have risen over the integrity of media and telecommunications, as a result of Chinese business transactions. Among these is the Huawei Technologies Company, which was detailed in chapter 2 as having been accused of a decade-long string of espionage incidents.

> Huawei has a major presence in Taiwan's telecommunications sector: It has secured contracts to supply FarEasTone with approximately $36 million worth of equipment and maintenance services for wireless network controllers and base stations and to supply Asia Pacific Telecom with approximately $683 million worth of 3.5G networking and communications equipment. It has also built Taiwan Mobile's fixed-line Ethernet network and manufactured many of the headsets marketed by Chunghwa Telecom. Furthermore, nearly all 3G mobile network cards used in Taiwan incorporate parts produced by Huawei. (USCC Annual Report 2012)

Additional incidents which have drawn government and public opposition include a billion-dollar deal to "acquire the cable TV services of China Network Systems," the construction of direct undersea cable links between Taiwan and China, and the broadcasting of Chinese radio stations (USCC Annual Report 2012). In addition to disabling service in a time of conflict, these communication tools could also be used to provide false information and conduct PSYOPs. From 2004 to 2014, there have been approximately 20 high-profile incidents of alleged Chinese traditional espionage against Taiwan, which directly affect cyber warfare capability (Recent Developments in China's Relations 2014). Recent cases include a Lenovo computer employee, an air force captain at a Taiwan radar surveillance center, and Major General Lo Hsien-Che, who is believed to have given the PRC "highly sensitive information regarding Taiwan military communications and command and control systems" (USCC Annual Report 2012). Increased interaction through the opening up of economic and cultural exchanges has allowed increased ability to conduct such activities. Ironically, China can even obtain some value out of espionage activities which have been exposed and resulted in arrests. By using IO, China can weaken the reputation Taiwan's national security industry. This will problematize military recruitment efforts and lower the perceived ability to defend against an invasion. It can also cause the United States to reconsider the amount and quality of its arms sales to Taiwan out of concern that these would be transferred to China.

Pre-war preparations can be used to map network architectures, catalogue zero-day exploits, and establish a covert presence within networks. In some cases, the identification of weak points in Taiwan's critical infrastructure does not require CNE and can be accomplished through IO. For example, natural disasters, such as an earthquake in 1999 and a typhoon in 2001, revealed weaknesses and "single-point failure nodes" in Taiwan's "telecommunications, electric power, and transportation infrastructure" (China's Proliferation Practices 2008). These weaknesses could be targeted through CNA, precision strikes, or physical sabotage. Further, a landslide revealed that the loss of a single power grid tower is capable of knocking out 90 percent of the power grid "in the central mountainous region" (China's Proliferation Practices 2008). The ability for this type of information to be used offensively in a conflict casts suspicions on other agreements, like Taiwan's "Straits Exchange Foundation" and China's "Association for Relations Across the Taiwan Strait," agreeing to cooperate "in the areas of earthquake monitoring and meteorology" (USCC Annual Report 2014). "Building information, including the location of the President's office," and daily activities, are "openly available on the Internet" (China's Proliferation Practices 2008). This is even more significant given the lack of security present during the 2004 assassination attempt on President Chen Shui-bian and Vice President Annette Lu (President and deputy survive assassination attempt 2004). Further, port statistics and detailed information on the movement of shipments are available online. IO-enabled planning can also yield IO tactics which could be used during conflict, such as "an internet rumour in 1999 that a Chinese Su-27 had shot down a Taiwan aircraft [which] caused the Taipei stock market to drop more than two percent in less than four hours" (China's Proliferation Practices 2008). In an effort to "win a battle without fighting," NCW has a minimal role due to its direct ties to military hardware. It does, however, maintain significance and overlap with the other branches. NCW provides a psychological and credible deterrent to conflict, and it is a primary recipient of the technology transfer and targeting intelligence obtained through IO and CNE. In the event of a conflict, its role dramatically increases.

LIMITED TOTAL WAR

If the Taiwan situation were to deteriorate into conflict, such as Taiwan declaring independence, it is unclear how severe China's application of force would be. At one end of the spectrum, China could apply the minimal amount of force necessary to return the situation to status quo. This seems unlikely given China's strong stance against independence and the resolve required by Taiwan to make such a declaration. At the other end of the spectrum, there is unrestricted warfare, which could result in the full application of U.S. force to intervene. Theoretical examples include China using ASAT weapons against U.S. satellites, EMP bursts over continental United States, CNA against financial systems, IO conducted over broadcast systems, and the full might of China's armed forces against Taiwan. The United States possesses equivalent capabilities, in addition to the ability to project power globally in the land, air, and sea domains, such as aerial bombardment of the Chinese mainland. As stated in the opening of this chapter's previous subsection, *Victory without War*, this is an unlikely escalation which all parties hope to avoid; the consequences are too high. Therefore, the most likely type of direct conflict to occur in a Taiwan contingency is an intermediate one. Even Net-Centric Warfare, which is directly tied to traditional military hardware, has been shown to possess characteristics of long-distance, precision, and reduced casualties.

China would consider all possibilities, including civilian and commercial targets, yet it would refrain from options which have exceedingly high consequences. Thus, it would be total, yet limited, warfare. According to Mark Stokes, current Chinese doctrine seeks "disruption and paralysis, not destruction"; and it draws on the teachings of Mao Zedong, who promoted fighting from a great distance, disrupting an enemy's senses, and attempting to confuse commanders (China's Proliferation Practices 2008). According to the U.S. Department of Defense 2012 Annual Report to Congress, the PLA's best option at present is the use of "limited force or coercive options" (U.S. Department of Defense 2012). This would entail limited computer and traditional attacks against key civilian, government, and military targets, along with the use of special operations forces, as opposed to a full-scale amphibious landing (Pilger 2015). The two main centers of gravity in such a conflict would be the will to fight on the part of Taiwan and the United States. China would seek to convince Taiwan to surrender as quickly as possible, and it would attempt to deter or delay U.S. involvement long enough for Taiwan to do so. In addition to surgical strikes, assassin's mace techniques would be suitable against Taiwan but not against the United States, as this could increase resolve to intervene or retaliate (Pillsbury 2004). In the case of the United States, China would undertake restrained actions in accordance with active defense. When applied to Taiwan, *The Science of Campaigns* and *The Science of Military Strategy* promote the targeting of C4ISR and logistics networks, calling them "the heart of information collection, control, and application on the battlefield" and "the nerve center of the entire battlefield" (U.S. Department of Defense 2013).

The balance of power in a China-Taiwan conflict has shifted in China's favor. China possesses approximately 1.25 million active-duty personnel, along with more than 2,000 combat aircraft and up to 280 naval vessels that could be used in a Taiwan conflict scenario (U.S. Department of Defense 2012; USCC Annual Report 2014). By comparison, Taiwan has roughly 130,000 active-duty personnel, "410 combat aircraft and 90 naval" ships (U.S. Department of Defense 2012; USCC Annual Report 2014). China's 1,200 mobile SRBMs can be fitted with a range of warheads, including cluster munition payloads, which are optimal for runway cratering (Recent Developments in China's Relations 2014; Taiwan-China: Recent Economic 2010; U.S. Department of Defense 2014). In the opening stages of a conflict, China could ground Taiwan's air force by targeting these runways. Between 2004 and 2014, the accuracy of these missiles has improved from around 300 meters to 20 meters. Taiwan's missile defense systems could only stop a small percentage of these due to the high volume of incoming missiles and the use of decoys. Follow-on strikes could target remaining aircraft, command and control centers, and ports, while anti-ship missiles could target vessels which were out to sea. Sea Lines of Communications are vital to Taiwan as it imports nearly 98 percent of its fossil fuel energy needs and lacks strategic reserves to last beyond two weeks (Hammond 2014). As detailed in chapter 4, China's modernization efforts have improved its ASBMs, early-warning aircraft, integrated air defense, ISR equipment, over-the-horizon radars, stealth aircraft, satellites, UAVs, and UCAVs.

NCW works best against an opponent with an intermediate level of NCW development, and Taiwan fits this category. Rigid elements of defense, like "fixed land-based coastal surveillance radars" and a "lack of mobile systems," combined with "insufficient infrastructure hardening," provide attractive targets for PLA precision strikes (Pilger 2015; USCC Annual Report 2013). Special operations forces, which have an increased role under NCW, can provide battle damage assessment, reconnaissance information for targeting, and sabotage, all of which are aided by the pre-conflict intelligence gathered through CNO and IO. Further, these forces can serve to overtake and hold key facilities with a reduced profile, such as Russia's alleged actions in the annexation of Crimea (Ash 2015). Obtaining entry into Taiwan prior to the conflict has been eased, due to "ethnic and linguistic homogeneity and the dramatic increases in cross-strait people flows" (China's Proliferation Practices 2008). Meanwhile, IO and CNO could be used to spread misinformation among the populace, such as falsified declarations of surrender, inaccurate war assessments, U.S. abandonment, or a popular sentiment to "lay down arms."

Depending on operational assessments, China may choose to use surgical strikes against key networks and nodes or take a scattershot approach. The former is more difficult to accomplish, but it minimizes damage. If China opts for maximum impact, it could begin disrupting, denying, degrading, or destroying as many networks as it is capable of, including critical nodes with a desire to cause cascading failures. The director of Taiwan's National Security Bureau stated in 2013 that Chinese CNA and CNE which traditionally target government networks had shifted focus toward "civilian think tanks, telecommunications service providers, Internet node facilities and traffic signal control systems" (Hsiao 2013). While there has been a lack of reported state-sponsored CNA, and computer networks encompass a great range of technical specifications, it remains a credible threat. Non-state actors have demonstrated a wide range of possibilities, and the 15-year span of CNE incidents detailed in chapter 2 show an even greater range of targets whose networks were capable of being breached. Maintaining a covert presence within networks prior to conflict, and spreading latent malware that contains a timer or can be remotely activated, can allow for a mass attack to be synchronized. At the start of conflict, and without prior notice, China could quickly mobilize full-scale online people's war by employing its large Internet population, militias, and botnets, guided by IO, PLA officers, and the leaders of hacker groups who could disseminate commands and automated toolkits. This would add quantity to quality. Web page defacements and DDoS could weaken Taiwan's morale and disrupt communication, possibly causing panic. If ICT services remain online, propaganda images and texts could be sent to mobile phones, and malware-laden e-mails could be sent purporting to contain links to breaking news on the war or government service announcements. Attacking financial networks and point-of-sale software would disrupt the ability of society to function. This could be coupled with tangible strikes on undersea fiber-optic cables, communication nodes, and transportation chokepoints. While chaos ensues in Taiwan under blackout conditions, China could use IO internationally to minimize negative perceptions and disguise the true nature of the situation. Domestically, China could censor negative information and propagandize the conflict.

In the event of a Taiwan conflict, China would use cyber warfare to delay U.S. involvement long enough for Taiwan to capitulate. At the same time, it would do so in a restrained fashion, if possible, to limit the U.S. response after reunification. One asymmetric means for accomplishing this is to target the logistics apparatus of U.S. force deployment, such as NIPRNet and USTRANSCOM (see chapter 2 for previous intrusions into these systems). This includes the organization of forces, food supplies, uniforms, and/or communication which are often organized through networks that are connected to the Internet. CNO could also delay re-supply to the region by misdirecting stores, fuel, and munitions, corrupting or deleting inventory files, and thereby hindering mission capability. The United States is at a geographic disadvantage in a Taiwan conflict. While Taiwan is roughly 180 kilometers from China, it is 2,780 kilometers from Guam, 8,150 kilometers from Hawaii, and 10,370 kilometers from the U.S. West Coast (Taiwan-China: Recent Economic 2010). If the PLA lacks the ability to find exploits in these networks, they could simply conduct DDoS attacks to bring them down long enough for a Taiwanese surrender. The United States' reliance on commercial entities for communications could also be targeted, and a significant disruption would tax a limited amount of IT personnel capable of administering fixes (Bronk 2011). Non-Internet military communications, including satellites, could also be disrupted through electronic jamming. By using proxies, China could use U.S. computers to conduct attacks on other states, or use the computers of other states to conduct attacks on the United States, in the hope that it will draw them into conflict with each other. If China is unable to delay the United States through such methods, or a strong U.S. presence was already in the theater of operations at the onset of hostilities, China may have to conduct limited kinetic strikes and forgo anonymity.

China could use anti-access area denial methods outlined in chapter 4 to conduct a blinding campaign against U.S. C4ISR and use anti-ship missiles to push U.S. aircraft carriers outside the First Island Chain. A less lethal method could be the use of non-nuclear EMP warheads to disable an aircraft carrier and serve as a warning shot (Thomas 2009, 224). Penetrating the on-board computer systems of aircraft and ships or activating latent kill switches and exploiting backdoors in military hardware microchips are all theoretically possible. These are more difficult objectives to achieve than A2/AD capabilities; however, allegations of Chinese CNE have shown a long-term and sustained effort to develop such a capability. Attention has been paid to accessing information on a specific system, like the F-35 fighter jet, and the concept of embedding exploits in microchips has been discussed by the PLA since 1988. Escalating the situation, extended-range precision strikes could target runways and key facilities at forward bases in Japan (Taiwan-China: Recent Economic 2010). This could precipitate a decision to expand operations and seize the Senkaku/Diaoyu islands, since many political lines would already have been crossed. Alternatively, China could abstain from kinetic strikes against the United States and Japan, and instead target computer networks in the United States. It is unlikely that severe disruptions to U.S. critical infrastructure could distract or delay military intervention, especially in a conflict which China would prefer to keep short, but these disruptions could serve as a deterrent. When combined with IO, it could also be used to turn popular U.S. opinion against involvement.

CYBER DEFENSE CONSIDERATIONS

Cyber warfare is not restricted to China, so Beijing must consider, and attempt to counter, Taiwan's capability in this field. Like China and the United States, Taiwan is developing its NCW capability (for greater detail on U.S. and Chinese NCW, see chapter 4). Taiwan has sought to improve its C4ISR and streamline forces, and to this end, it has introduced "one of the world's most sophisticated advanced tactical data link networks" (Taiwan-China: Recent Economic 2010; Pillsbury 2004). As of 2014, this system remains in the early stages of development. Although, these data links are a significant step toward interconnecting command, platforms, sensors, and shooters that will provide enhanced situational awareness and survivability. Adding to this, the United States approved arms sales to Taiwan worth over $12 billion in the period 2010–2014 since 2010, including upgrades to Taiwan's F-16 A/B fighter jets (U.S. Department of Defense 2014; USCC Annual Report 2014). These upgrades comprise a number of items which are frequently mentioned in Net-Centric Warfare literature, such as active electronically scanned array radars, data link terminals, GPS navigation equipment, helmet targeting systems, improved communication equipment, improved electronic warfare systems, laser-guided munitions, logistical support, night-vision systems, and updated cockpit computer systems (USCC Annual Report 2011). Taiwan also operates variants of the U.S. Lockheed P-3 Orion reconnaissance aircraft. The One-China Policy limits Taiwan's options for the purchase of weapons; however, it is domestically developing long-range anti-ship cruise missiles, stealth technology, UAVs, and UCAVs (USCC Annual Report 2014). Collaboration between Taiwan and the United States could increase, particularly under administration changes and as the U.S. rebalance to Asia policy evolves. Taiwan is located in an advantageous position for enhanced ISR against China and shares common security and political views with the United States.

Similar to China, yet at an earlier stage of development, Taiwan can develop A2/AD capabilities until its NCW competency matures. Taiwan's 2013 *Quadrennial Defense Review* calls for "fielding innovative and asymmetric capabilities" (USCC Annual Report 2013). Among these are truck-mounted cruise missiles which utilize commercial radar and anti-radiation systems for target acquisition (Recent Developments in China's Relations 2014). Taiwan's expanding patrol fleet is another example of asymmetric capability. These ships incorporate stealth technology and are armed with anti-ship cruise missiles, AV-2 decoy launchers, and electronic warfare support (Murray 2013; Pilger 2015). In comparison to some of the most advanced NCW systems, these are relatively inexpensive, allowing for greater numbers to be fielded, and their mobility will challenge the extent of China's speed and precision targeting. Additional efforts are underway to harden fixed targets and limit the PLA's ability to deny Taiwan use of its runways. For the latter, Taiwan is improving its "rapid runway repair," reducing the take-off weight of aircraft to limit the required length of runways, and exploring the option of highway landings (Recent Developments in China's Relations 2014). While China's arsenal of SRBMs is imposing, it also contains potential asymmetric weaknesses by virtue of being a networked system, and Taiwan has reportedly identified single-point failures within this system (Taiwan-China: Recent Economic 2010; USCC Annual Report 2013). Even if Taiwan cannot defeat a Chinese invasion, its goal is to raise the cost of such an attempt to an unacceptable level, or delay long enough for U.S. assistance. In this regard, the inability of Taiwan's runways to sustain an air campaign, or the inability of Taiwan's missile defense systems to stop all incoming missiles, does not mean such pursuits should be abandoned; they raise the stakes in China's calculations. Further, continued pursuit of these systems advances Taiwan's long-term NCW capabilities.

Taiwan is frequently targeted by alleged Chinese CNO and has endured such intrusions for more than a decade (Hsiao 2013). As a result, Taiwan's computer network defense has strengthened. It established one of the world's first "cyber operations commands in 1999" and has operated cyber units since the mid-1990s (Recent Developments in China's Relations 2014). Taiwan has a thriving computer security industry and possesses access to China's microchip supply chains; therefore, like China, Taiwan could explore the use of embedded exploits to corrupt systems. China's increasing reliance on computer networks and ties to Taiwan provide new avenues for attack. Taipei-based corporations that employ or handle the financial transactions of Chinese workers in the mainland, for example, could cease communications with China in time of conflict, thereby preventing millions of Chinese workers from receiving their income deposits (Recent Developments in China's Relations 2014). Taiwan's Information and Electronic Warfare Command is comprised of 3,000 military officers (Hsiao 2013). As of 2014, Taiwan is increasing its cyber warfare budget, bolstering the defense of C4ISR networks, increasing inspections, and increasing the training of personnel (USCC Annual Report 2014). A fourth cyber unit was also established within the Ministry of National Defence, and a cyber warfare simulation facility was constructed to test the defenses of Taiwan's civilian critical infrastructure (USCC Annual Report 2013). China allegedly tests its CNO capabilities against Taiwanese systems and uses Taiwan computer networks as a staging ground for conducting attacks against the United States (Gold 2013). This mutual threat fosters Taiwan-U.S. collaboration, and Taiwan has sought participation in U.S. cyber exercises (Gold and Wu 2015). The United States can gain valuable knowledge from Taiwan due to its experience in dealing with Chinese intrusions and its shared language and cultural understanding.

Chinese cyber strategies have the potential to backfire. China's quasi-endorsement or tolerance of patriotic hackers, for instance, could result in these groups undermining an operation such as conflict with Taiwan (China's Proliferation Practices 2008). Information operations in particular can have unintended influence. The Sunflower Student Movement demonstrates the level of mistrust China must overcome, yet overly friendly messages intended to boost soft power might be viewed as propaganda. China cannot remove the threat of force or Taiwan would de-clare independence, yet having 1,200 SRBMs across the narrow strait makes it difficult to establish trust and a favorable relationship. Taiwan or the United States could conduct their own IO to heighten this threat perception as a way to achieve public support for increased defense spending. China must compete with the globally produced messages reaching Taiwan; there is no Golden Shield to aid in this task. Develop-ments in Hong Kong, it should be noted, are closely followed by Taiwa-nese people and impact negatively on the "one country, two systems" principle (USCC Annual Report 2014).

CONCLUSION

Cyber warfare occupies a central role in China's attempts to reunify with Taiwan, and it can be applied flexibly. During peacetime, information operations are key, as China is attempting to influence Taiwan's ac-tions—including a relinquishing of media control, establishing favorable public opinion, and the fostering of greater dependence. At the same time, computer network exploitation is used as a contingency to prepare for conflict. In a sense, the keys to virtual doors are obtained and virtual explosives are planted. Online espionage also provides NCW technology transfer, while NCW itself is relegated to a deterrent role. This is China's preferred strategy. However, if conflict becomes necessary, the applica-tion of cyber warfare adapts. IO remains important, but its significance is eclipsed by the then dominant NCW and computer network attack. Cy-ber warfare is also applied differently to the two opponents in this sce-nario. For the United States, China seeks to delay and deter with minimal damage inflicted. For Taiwan, China prefers key-point strikes, but the speed of the operation is paramount. The primary goal is Taiwanese surrender and reunification, so a scattershot approach may be permitted depending on capability and assurance. This breakdown in the applica-tion of cyber warfare to the Taiwan issue does not apply equally in all situations. It is a fluid formula to account for the high number of vari-ables in international relations. The tripartite of cyber warfare remains conducive to China's goals and culture, while offering a holistic approach and creating synergy.

NOTES

1. Taiwan's largest trading partners in 2013 were China, Hong Kong, Japan, and the United States, at 22, 7, 11, and 10 percent, respectively (USCC Annual Report 2014). This places Taiwan in a difficult position with its largest trading partners being roughly split between Chinese and U.S. policy interests.

2. The two steps forward are considered a bold move; and despite backlash, this is allowed to set. Eventually, one step back is conceded to ease tension, yet it remains a net gain of one step. Examples include China's establishment of an Air Defence Identification Zone (ADIZ) in the East China Sea, and the placement of an oil rig in waters claimed by Vietnam (see Keck 2013; Vuving 2014).

Conclusion

Through a conceptual framework and historical analysis, this research has addressed gaps in China's strategic doctrine and the field of cyber warfare. A lack of transparency in Chinese military affairs, the infancy of the field of cyber warfare, and unconsolidated cyber terminology contributed to the absence of China's cyber warfare doctrine in open source material. These problems were addressed through a comprehensive literature review and the development of an all-inclusive terminology model, based on key traits and majority patterns. Chapter 1 of this book found that computer network operations aid China's modernization goals, and it corresponds to China's historical tendency toward asymmetric warfare. The PLA acknowledged the existence of offensive cyber units to international audiences in 2011 and 2013, although PLA literature places their creation one decade prior. It was found that China maintains at least 20 cyber units, 33 CNO militia, and 250 hacker groups. Key entities responsible for the development of CNO are the General Staff Department's Third and Fourth Departments. Computer network attack is handled by the Fourth Department, while exploitation and defense are the responsibility of the Third Department. A collection of additional departments and ministries within the Central Military Commission and the State Council provide supplementary activities. All computers which are connected to the Internet are susceptible to hacking, and intruders can obtain up to complete control, therefore the range of targets and types of theoretical tactics are immense. Networks with unique architectures, such as industrial control systems, mobile phones, point-of-sale systems, and satellites, are also vulnerable to attack and exploitation. Civilian networks are particularly vulnerable as the perceived risk must outweigh the cost of implementing advanced security measures.

As of 2016, there have been few state-sponsored computer network attacks. However, criminal organizations, hacker groups, and other non-state actors have demonstrated that attacks are possible. Documented examples include attacks on electrical grids, financial systems, and transportation networks. In contrast, computer network exploitation seeks to remain covert. Global reach and a high degree of anonymity make CNE ideal for the theft of intellectual property, although it also provides the ability to eavesdrop and set up for an attack. One way to prepare for a potential conflict is the use of embedded backdoors, and PLA theorists have been researching this concept since the late 1980s. With the world's largest Internet population, and second largest economy, China is frequently the target of computer attack and exploitation by non-state actors. Additionally, as of 2016, there are over 100 states developing CNO capabilities. There have been three recorded incidents in which non-state actors have targeted Chinese industrial control systems, and four incidents of state-sponsored exploitation targeting large Chinese information gateways. These are in addition to thousands of smaller attacks recorded each year by China's Computer Network Emergency Response Technical Team. Further, the widespread media attention given to alleged Chinese CNE may make Chinese computers a preferred choice for use as proxies in attacks by other states. China's negative reputation could add to the belief that China was the culprit.

Chapter 2 traced the origins of PLA cyber units to the founding of hacker groups between the years 1998 and 2002. Nationalism served as a common bond between these groups, and they were involved in large-scale conflicts with Indonesia, Japan, Taiwan, and the United States. These groups were not state-sponsored, and their primary mode of operations was attack, yet they are credited with demonstrating the Internet's emerging military application. Further, their computer expertise made them candidates for recruitment by the PLA. Allegations of Chinese state-sponsored CNE gained cohesion around 2002 and increased steadily throughout 2016. This research identified and analyzed 19 advanced persistent threats and over 100 individual CNE incidents which were attributed to China based on six attribution factors. Intrusions spanned 71 countries and targeted the full spectrum of industries, from critical infrastructure and corporations, to government and military networks (see appendix A). Online espionage allows instantaneous global reach, plausible deniability, and minimal repercussions. Among these incidents there were recurring themes, such as allegations of implanting backdoors in microchips, particularly by the Huawei Technologies Company and the computer manufacturer Lenovo. NASA and information pertaining to the F-35 fighter aircraft were frequent targets, and there was an alleged pattern of using traditional espionage to facilitate online espionage. Between 2010 and 2014, international condemnation of alleged Chinese exploitation experienced a sharp increase. Open source material from this period contains a greater amount of evidence linking intrusions to state-sponsorship and stronger stances in assigning blame. This four-year span was also marked by high-profile withdrawals of government and commercial contracts with China, and increased legal action taken against accused perpetrators. Despite these counteractions, there has been no reduction in alleged intrusions.

In chapter 3, information operations was found to occupy a large role in Chinese strategy, despite being the only legal[1] (non-hacking) branch of cyber warfare. The Three Warfares was found to be an inadequate way of conceptualizing IO, and it was replaced by the concept of online messages influencing actions. This encompasses Internet censorship, media, propaganda, and psychological operations, and it can target domestic (internal) or foreign (external) audiences. Further, IO can be utilized during peace or war. Information operations exploitation uses false, misleading, or true messages to cause a desired action, while information operations defense attempts to block, change, or counter undesirable messages. These messages can be transmitted through a range of mediums, such as commentary, images, news, or video. A number of tactics were revealed for promoting or demoting messages, such as drowning out negative comments, blocking specific languages, and artificially inflating polls. In comparison to other states, China exerts a high level of control over domestic Internet activities. This is epitomized by the Golden Shield, which utilizes a multilayered approach to censorship. Information being transmitted to China is first filtered at large international hubs and subsequently scrutinized by an array of individuals and software programs. An array of government entities was identified as fostering the development of Chinese IO dating back to the mid-1990s. These organizations were spread throughout the Central Military Commission and State Council, and efforts are underway to streamline this apparatus. China utilizes thousands of paid commentators, known as the 50 Cent Party, to boost public support of government policies and improve China's image. Defensively, China must prevent foreign and domestic messages from revealing state secrets, threatening national security, or causing social unrest.

Chapter 4 examined the historical and linguistic ties which interconnect Net-Centric Warfare with the Revolution in Military Affairs, Electronic Warfare, Anti-Access Area Denial, and Informationization. By consolidating these core concepts on the future of warfare, an improved understanding of Chinese NCW development was obtained. NCW was revealed to be largely synonymous with RMA and informationization, and it possesses the most descriptive and precise term among the three. Electronic warfare has been identified as a subcomponent of NCW, and A2/AD is the use of NCW under set conditions. Net-Centric Warfare rose to prominence in the 1990s following U.S. technologic dominance in the Gulf War. NCW refers to the linking of command, platforms, sensors, and shooters through computer networks. Advanced military hardware relies on computerized systems and networks, and NCW seeks to link all of these systems together, within a "system of systems." This allows enhanced situational awareness, increased speed, increased precision, and the ability to act from long distances with smaller forces. In short, it increases overall battlefield effectiveness and efficiency. NCW is not only the application of advanced technologies and networked systems, it is changes in strategy, organization, and training which allow a military to take full advantage of their potential. A recurring theme is the proliferation of technology, which is allowing less advanced militaries to rapidly improve their capability, as well as adopt asymmetric tactics to place NCW-dominant militaries at risk. Identified drawbacks to implementing NCW include a high financial cost, the rapid pace of technologic change, and the introduction of new vulnerabilities, such as information overload and an inability to function in the event of a network disruption. As of 2016, China possesses a regional NCW capability and is rapidly advancing.

Chapter 5 applied China's cyber warfare capability to the Taiwan issue. This revealed two primary operational strategies. The first is to attain reunification with Taiwan without conflict. This entails the use of information operations to boost China's image among the Taiwanese population. IO is also used to increase cross-strait economic and cultural integration, deter U.S. arms sales, and empower pro-reunification Taiwanese politicians and business owners. At the same time, CNE is used to prepare for the possibility of conflict. This includes embedding backdoors in microchips, mapping networks, and maintaining a covert presence within networks. CNE can also yield technology transfer to boost China's NCW capability, and this capability can be used as a deterrent to conflict. If the Taiwan issue were to deteriorate, and conflict was deemed necessary, China would change its operational procedure. China would attempt to obtain Taiwan's surrender before the United States could intervene. NCW and CNA would be used to target critical nodes within Taiwan, such as C4ISR, runways, and civilian communication and transport hubs. Special Forces would also be utilized, rather than a full-scale amphibious invasion. Key-point strikes and Special Forces are characteristic of NCW, which places an emphasis on long-distance precision strikes, speed, and streamlined forces. China would take a more restrained approach in regard to U.S. forces, in order to minimize consequences. This includes targeting C4ISR and the logistical aspects of force deployment by using NCW and CNE, respectively. CNA against U.S. civilian computer networks, such as financial systems, could act as a deterrent. Similarly, if necessary, anti-ship missiles, jamming, and precision strikes on forward bases could be used to push the U.S. fleet outside of the First Island Chain. During conflict IO would be used to weaken enemy morale and cause confusion or fear. Beyond opponents, IO can be used to minimize negative perceptions of the conflict and win support. Cyber warfare suits China's culture and offers an enhanced, and fluid, means to attain its goals.

The foregoing chapters of this book have provided four significant contributions to the field of study and articulated an emerging Chinese doctrine of cyber warfare. First, a clear conceptual model for cyber warfare terminology has been created. Second, government organizations responsible for the development of cyber warfare have been identified and visually mapped. Third, a complete list of open source allegations of Chinese CNE has been assembled and analyzed. Fourth, a comprehensive review and historical analysis of China's cyber warfare development has been conducted. These four contributions did not exist in prior open source knowledge, let alone in such depth and focus.

There are several key trends present in China's cyber warfare doctrine. First, China takes a holistic approach. This is evident in the three-branch terminology model which was developed in accordance with the literature, and evident in their willingness to target civilian and non-military targets, which in some cases is less destructive. Second, there is a focus on attacking critical nodes. In CNO this comprises information hubs, critical infrastructure, and nodes which can cause cascading failures or disrupt the will of an opposing force. Examples of key-point strikes in NCW are C4ISR and logistics. China also maintains a scatter-shot approach as well, as shown in online people's war. Reflective of China's past, Sun Zi's deception flourishes in cyber warfare. This is exemplified by China's alleged preference toward CNE to leapfrog in modernization and use of information operations to impact on an opponent's will to fight, rather than resorting to direct conflict. China is pursuing asymmetric deterrents and utilizing CNE in the short term, while building toward regional dominance and securing global interests. A final identified pattern is China's preference for spear phishing as a low-level means of breaching computer networks.

Supporting the above observations is a body of literature that suggests China is pursuing a more streamlined structure for cyber warfare organization and development. This was evidenced by the transition from INEW to information confrontation, and the renaming of the Internet Security and Informationization Leading Group. By using the three-branch model proposed in this book, which is based on architectural distinctions, China (among other states) could consolidate terminology and collaboratively develop new concepts based on a shared model, rather than developing competing and overlapping concepts (see figure 0.1; table 0.1). Additional solutions were revealed in the identification and visual mapping of key government entities responsible for the development of cyber warfare (see figure 1.1). Currently, the 3PLA has responsibility for CNE and CND, while CNA is assigned to the 4PLA. However, the three components of computer network operations utilize the same skillset and equipment. Additionally, a hacker can transition seamlessly from exploitation to attack. Therefore, it would be more productive to place all CNO components under the same authority. Secondly, EW is separate from the Department of Informationization, yet these too would utilize the same equipment and skills. Lastly, IO has the most dispersed authority of all three branches. Consolidating these activities through the GPD and SCIO appears to be the optimal choice.

The increasing international condemnation over alleged Chinese computer network exploitation between the years of 2010 and 2014, along with the over 100 incidents and 19 APTs revealed in chapter 2, have implications for policy and practice. If these allegations are true, China should attempt to improve its CNE stealth capability. In either case, subtle application of IO could be used to help repair China's reputation in this regard.

Three recommendations for further research emerge from this book. First, research could be conducted to provide greater detail on the technical specifications of NCW equipment. In particular, it could seek to identify the degree to which these systems are different from one another, and whether there are sets of hacking methods to which a great number of systems are vulnerable. Second, an investigation into semi-closed networks could reveal the degree to which these networks are separated from the Internet. Subsequently, Figure 0.2: Venn of Cyber Warfare with Select Callouts could be adjusted to reflect more accurately the revised understanding. Lastly, the application of Chinese cyber warfare demonstrated in chapter 5 could be applied to other situations, such as a potential conflict with Japan over the Senkaku/Diaoyu islands. Application to multiple scenarios might reveal trends which are present in a majority of situations, or possibly, common to the application of cyber warfare as a whole, not only China.

The literature contains no comprehensive and clear view of China's cyber warfare doctrine, because China has never published one. This research has shown a holistic conceptual and historical approach can reveal more than the literature suggests. In doing so, it has established a foundation stone for scholarly research on a topic of profound significance to China's wider strategic doctrine.

NOTE

1. This is legal at a state level; it does not apply to individuals who use IO to aid criminal activity or cause social disturbances.

Appendix A

Countries Allegedly Exploited by China and Supporting References

- Afghanistan (Kaspersky Lab 2013a)
- Armenia (Villeneuve and Sancho 2011)
- Australia (China blamed after ASIO 2013; China says hit 2011; Doherty, Gegeny, Spasojevic, and Baltazar 2013; Kaspersky Lab 2013a; Krekel 2009; ONCIX 2011; Tkacik 2008; USCC Annual Report 2012; USCC Annual Report 2013; Wilkinson 2010)
- Austria (Kaspersky Lab 2013a; Kaspersky Lab 2013b; Villeneuve and Sancho 2011)
- Azerbaijan (Villeneuve and Sancho 2011)
- Bangladesh (Information Warfare Monitor 2009; Kaspersky Lab 2013a)
- Barbados (Information Warfare Monitor 2009)
- Belarus (Kaspersky Lab 2013a; Kaspersky Lab 2013b; Villeneuve and Sancho 2011)
- Belgium (Chien and O'Gorman 2011; Information Warfare Monitor 2009; Kaspersky Lab 2013a; Luard 2005; Mandiant 2013; U.S. Department of Defense 2009)
- Bhutan (Information Warfare Monitor 2009)
- Brazil (Novetta 2014)
- Brunei (Information Warfare Monitor 2009)
- Cambodia (Kaspersky Lab 2013a; Villeneuve and Sancho 2011)
- Canada (Alperovitch 2011; Doherty, Gegeny, Spasojevic, and Baltazar 2013; Information Warfare Monitor 2010; Kaspersky Lab 2013a; Tkacik 2008; Mandiant 2013; Nortel hit by suspected Chinese 2012; USCC Annual Report 2012)
- Chile (Kaspersky Lab 2013a)
- Cyprus (Information Warfare Monitor 2009)
- Czech Republic (Villeneuve and Sancho 2011)
- Denmark (Alperovitch 2011; Chien and O'Gorman 2011)
- France (Committee on Foreign Affairs 2011; Doherty, Gegeny, Spasojevic, and Baltazar 2013; Jowitt 2013; Kaspersky Lab 2013b; Krekel 2009; Luard 2005; Mandiant 2013; ONCIX 2011; Tkacik 2008; Villeneuve and Sancho 2011)

- Germany (Alperovitch 2011; Digital Spying Burdens 2013; Doherty, Gegeny, Spasojevic, and Baltazar 2013; Glanz and Markoff 2010; Information Warfare Monitor 2009; Jowitt 2013; Kaspersky Lab 2013a; Kaspersky Lab 2013b; Krekel 2009; Tkacik 2008; USCC Annual Report 2012; Villeneuve and Sancho 2011)
- Greece (Kaspersky Lab 2013a)
- India (Alperovitch 2011; Doherty, Gegeny, Spasojevic, and Baltazar 2013; Information Warfare Monitor 2009; Information Warfare Monitor 2010; Kaspersky Lab 2013a; Luckycat Redux 2012; Mandiant 2013; Tkacik 2008; U.S. Department of Defense 2009; USCC Annual Report 2012; Villeneuve and Sancho 2011)
- Indonesia (Alperovitch 2011; Honker Union of China 2010; Information Warfare Monitor 2009; Kaspersky Lab 2013a; Novetta 2014)
- Iran (Information Warfare Monitor 2009; Kaspersky Lab 2013a)
- Ireland (Doherty, Gegeny, Spasojevic, and Baltazar 2013)
- Israel (Mandiant 2013)
- Italy (Chien and O'Gorman 2011; Kaspersky Lab 2013b)
- Japan (Alperovitch 2011; Chien and O'Gorman 2011; Doherty, Gegeny, Spasojevic, and Baltazar 2013; Honker Union of China 2010; Kaspersky Lab 2013a; Kaspersky Lab 2013b; Krekel 2009; Luckycat Redux 2012; Mandiant 2013; Novetta 2014; Raff 2013; USCC Annual Report 2012)
- Jordan (Kaspersky Lab 2013a)
- Kazakhstan (Kaspersky Lab 2013a; Villeneuve and Sancho 2011)
- Kyrgyzstan (Kaspersky Lab 2013a; Villeneuve and Sancho 2011)
- Latvia (Information Warfare Monitor 2009)
- Lithuania (Kaspersky Lab 2013a)
- Luxembourg (Mandiant 2013)
- Malaysia (Doherty, Gegeny, Spasojevic, and Baltazar 2013; Kaspersky Lab 2013a; Kaspersky Lab 2013b; USCC Annual Report 2013)
- Maldives (Kaspersky Lab 2013b)
- Malta (Information Warfare Monitor 2009)
- Mongolia (Kaspersky Lab 2013a; Villeneuve and Sancho 2011)
- Morocco (Kaspersky Lab 2013a)
- Nepal (Kaspersky Lab 2013a)
- Netherlands (Chien and O'Gorman 2011; Kaspersky Lab 2013b; Villeneuve and Sancho 2011)
- New Zealand (Krekel 2009; Tkacik 2008)
- Norway (Mandiant 2013)
- Pakistan (Information Warfare Monitor 2009; Information Warfare Monitor 2010; Kaspersky Lab 2013a; Kaspersky Lab 2013b; Villeneuve and Sancho 2011)
- Philippines (Information Warfare Monitor 2009; Krekel 2009; Passeri 2012)
- Portugal (Information Warfare Monitor 2009)

- Qatar (Kaspersky Lab 2013a)
- Romania (Information Warfare Monitor 2009)
- Russia (Doherty, Gegeny, Spasojevic, and Baltazar 2013; Kaspersky Lab 2013a; Kaspersky Lab 2013b; Villeneuve and Sancho 2011)
- Saudi Arabia (Chien and O'Gorman 2011)
- Singapore (Alperovitch 2011; Doherty, Gegeny, Spasojevic, and Baltazar 2013; Kaspersky Lab 2013b; Mandiant 2013)
- Slovenia (Kaspersky Lab 2013a)
- South Africa (Mandiant 2013)
- South Korea (Alperovitch 2011; Doherty, Gegeny, Spasojevic, and Baltazar 2013; Information Warfare Monitor 2009; Kaspersky Lab 2013a; Kaspersky Lab 2013b; Krekel 2009; Novetta 2014; ONCIX 2011; Villeneuve and Sancho 2011)
- Spain (Villeneuve and Sancho 2011)
- Sri Lanka (Kaspersky Lab 2013b)
- Suriname (Kaspersky Lab 2013a)
- Sweden (Luard 2005)
- Switzerland (Alperovitch 2011; Mandiant 2013; Villeneuve and Sancho 2011)
- Syria (Kaspersky Lab 2013a)
- Taiwan (Alperovitch 2011; China says hit 2011; Doherty, Gegeny, Spasojevic, and Baltazar 2013; Honker Union of China 2010; Hsiao 2013; Information Warfare Monitor 2009; Information Warfare Monitor 2010; Kaspersky Lab 2013b; Krekel 2009; Mandiant 2013; Novetta 2014)
- Tajikistan (Kaspersky Lab 2013a; Villeneuve and Sancho 2011)
- Thailand (Information Warfare Monitor 2009; Kaspersky Lab 2013a; USCC Annual Report 2012; Villeneuve and Sancho 2011)
- Turkey (Kaspersky Lab 2013a)
- Turkmenistan (Kaspersky Lab 2013a; Villeneuve and Sancho 2011)
- UAE (Mandiant 2013; Villeneuve and Sancho 2011)
- Ukraine (Doherty, Gegeny, Spasojevic, and Baltazar 2013; Kaspersky Lab 2013a; Villeneuve and Sancho 2011)
- UK (Chien and O'Gorman 2011; Committee on Foreign Affairs 2011; Doherty, Gegeny, Spasojevic, and Baltazar 2013; Jowitt 2013; Kaspersky Lab 2013a; Mandiant 2013; Norton-Taylor 2007; ONCIX 2011; Protalinski 2012; Tkacik 2008; USCC Annual Report 2012; Villeneuve and Sancho 2011)
- United States (Information Warfare Monitor 2010; Mandiant 2013; Novetta 2014; U.S. Department of Defense 2014; USCC Annual Report 2013; et al.)
- Uzbekistan (Kaspersky Lab 2013a; Villeneuve and Sancho 2011)
- Vietnam (Alperovitch 2011; Information Warfare Monitor 2009; Information Warfare Monitor 2010; Villeneuve and Sancho 2011)

Appendix B

Alleged City or Province within
China responsible for CNE

Location	Possible PLA Unit	Major Campaigns	Sources
Beijing	66407, 61580, or 61046	Axiom; Ephemeral Hydra; Hidden Lynx; Operation Aurora; Operation DeputyDog	Clayton 2012; Digital Spying Burdens 2013; Krekel, Adams, and Bakos 2012; Stokes, Lin, and Hsiao 2011; USCC Annual Report 2011
Chengdu	78006	The Shadow network	Grow and Hosenball 2011; Information Warfare Monitor 2010; Krekel, Adams, and Bakos 2012; Stokes, Lin, and Hsiao 2011
Fujian P.	61716		Krekel 2009; Stokes, Lin, and Hsiao 2011
Guangzhou	75770	Code Red Worms; Titan Rain	Digital Spying Burdens 2013; Henderson 2007a; Krekel, Adams, and Bakos 2012; Stokes, Lin, and Hsiao 2011; Tkacik 2008
Hainan P.	Unknown	GhostNet	Information Warfare Monitor 2009
Hubei P.	Unknown		Krekel 2009
Jinan	72959	Fake Mandiant Report	Krekel, Adams, and Bakos 2012; Raff 2013; Stokes, Lin, and Hsiao 2011; USCC Annual Report 2011
Nanjing	73610		Krekel, Adams, and Bakos 2012; Stokes, Lin, and Hsiao 2011; USCC Annual Report 2012
Shanghai	61398	APT1; Shady Rat	Digital Spying Burdens 2013; Grow and Hosenball 2011; Mandiant 2013; Riley and Dune 2012; Stokes, Lin, and Hsiao 2011
Shanghai	61486		CrowdStrike Intelligence Report 2014; Stokes, Lin, and Hsiao 2011
Shaoxing	Unknown		Nusca 2010
Tianjin	Unknown	Myfip	Brenner 2005; Henderson 2007a
Wuhan	61726		Hsiao 2013; Stokes, Lin, and Hsiao 2011

Zigong	Unknown	Saporito and Lewis 2013

These are often stated as "from sites in x," "x-based IP addresses," or "server(s) located in x." Possible connections to specific PLA units are listed in column two. APT/campaigns (or select incidents) are noted in column three. Campaigns not listed, such as Avocado, Nitro, and Sykipot, are alleged to be from an unknown location in China. This list is restricted to only those incidents discussed in chapter 2.

Bibliography

Abagnale, Frank W. (2000). *Catch Me If You Can*. New York: Broadway Books.

Abrams, Marshall, and Weiss, Joe. (2008). Malicious Control System Cyber Security Attack Case Study–Maroochy Water Services, Australia. Retrieved on November 21, 2014, from http://csrc.nist.gov/groups/SMA/fisma/ics/documents/Maroochy-Water-Services-Case-Study_report.pdf.

Access to Information and Media Control in the People's Republic of China. (2008). Hearing before the U.S.-China Economic and Security Review Commission. Retrieved on September 15, 2014, from http://origin.www.uscc.gov/sites/default/files/transcripts/6.18.08HearingTranscript.pdf.

Adee, Sally. (2008). The Hunt for the Kill Switch. Retrieved on December 2, 2014, from http://spectrum.ieee.org/semiconductors/design/the-hunt-for-the-kill-switch.

Ahvenainen, Sakari. (2003). Backgrounds and Principles of Network-Centric Warfare. Retrieved on March 2, 2015, from http://www.csl.army.mil/SLET/mccd/CyberSpacePubs/NCW%20Background%20Principles.pdf.

Alberts, David S., Garstka, John J., and Stein, Frederick P. (2000). Network Centric Warfare: Developing and Leveraging Information Superiority. Retrieved on October 7, 2014, from http://www.dodccrp.org/files/Alberts_NCW.pdf.

Alperovitch, Dmitri. (2011). Revealed: Operation Shady RAT. Retrieved on October 2, 2014, from http://www.mcafee.com/us/resources/white-papers/wp-operation-shady-rat.pdf.

Amoroso, Edward E. (2011). *Cyber Attacks: Protecting National Infrastructure*. Burlington, MA: Elsevier, Butterworth-Heinermann.

Anderson, Eric C., and Engstrom, Jeffrey G. (2009). China's Use of Perception Management and Strategic Deception. Retrieved on September 15, 2014, from http://origin.www.uscc.gov/sites/default/files/Research/ApprovedFINALSAICStrategicDeceptionPaperRevisedDraft06Nov2009.pdf.

Andress, Jason, and Winterfeld, Steve. (2011). *Cyber Warfare: Techniques, Tactics and Tools for Security Practitioners*. Waltham, MA: Elsevier, Syngress.

Anonymous says it hacked Chinese government sites. (2012). Retrieved on October 3, 2014, from http://news.yahoo.com/anonymous-says-hacked-chinese-government-sites-040013396.html.

Anonymous says it will hack more Chinese sites. (2012). Retrieved on October 3, 2014, from http://www.spacedaily.com/reports/Anonymous_says_it_will_hack_more_Chinese_sites_999.html.

APCERT Annual Report. (2009). Retrieved on November 18, 2014, from http://www.apcert.org/documents/pdf/APCERT_Annual_Report_2009.pdf.

———. (2011). Retrieved on November 18, 2014, from http://www.apcert.org/documents/pdf/APCERT_Annual_Report_2011.pdf.

Areddy, James T. (2010). People's Republic of Hacking: "Panda" Exploit Offers Rare Inside Look at China's Cybercrime Networks. Retrieved on December 5, 2014, from http://online.wsj.com/articles/SB10001424052748704140104575057490343183782.

Aroor, Shiv. (2006). From Sky, See How China Builds Model of Indian Border 2400 km Away. Retrieved on September 22, 2014, from http://archive.indianexpress.com/news/from-sky-see-how-china-builds-model-of-indian-border-2400-km-away/9972/

Arrillaga, Pauline. (2011). AP Impact: China's spying seeks secret US info. Retrieved on October 3, 2014, from http://www.boston.com/news/nation/articles/2011/05/07/ap_impact_chinas_spying_seeks_secret_us_info/?page=full.

Arthur, Charles. (2011). China "targeted 48 chemical and military companies in hacking attack." Retrieved on October 7, 2014, from http://www.theguardian.com/technology/2011/nov/01/china-hacking-chemical-military-companies.

————. (2014). US accusations of Chinese hacking point to eight-year spying campaign. Retrieved on October 7, 2014, from http://www.theguardian.com/technology/2014/may/19/us-accusations-chinese-hacking-eight-years.

Ash, Lucy. (2015). How Russia outfoxes its enemies. Retrieved on May 1, 2015, from http://www.bbc.com/news/magazine-31020283.

Baldor, Lolita. (2013). US looking at action against China cyberattacks. Retrieved on October 5, 2014, from http://news.yahoo.com/us-looking-action-against-china-cyberattacks-223446697.html.

Barker, Garry. (2002). Cyber terrorism a mouse-click away. Retrieved on November 20, 2014, from http://www.theage.com.au/articles/2002/07/07/1025667089019.html.

Beech, Hannah. (2011). Meet China's Newest Soldiers: An Online Blue Army. Retrieved on November 23, 2014, from http://world.time.com/2011/05/27/meet-chinas-newest-soldiers-an-online-blue-army/.

Bitzinger, Richard A. (2014). Strategic ambiguity a hazard for Asian security. Retrieved on April 1, 2015, from http://www.eastasiaforum.org/2014/04/04/strategic-ambiguity-a-hazard-for-asian-security/.

Bonner, Raymond, and Spolar, Christine. (2013). Death in Singapore. Retrieved on November 23, 2014, from http://www.ft.com/cms/s/2/afbddb44–7640–11e2–8eb6–00144feabdc0.html#axzz2LG4u1Uph.

Boyd, Clark. (2008). Profile: Gary McKinnon. Retrieved on November 20, 2014, from http://news.bbc.co.uk/2/hi/technology/4715612.stm.

Branigan, Tania. (2010a). "Iranian" hackers paralyse Chinese search engine Baidu. Retrieved on November 1, 2014, from http://www.theguardian.com/technology/2010/jan/12/iranian-hackers-chinese-search-engine.

————. (2010b). US software firm sues China over Green Dam piracy. Retrieved on October 13, 2014, from http://www.theguardian.com/technology/2010/jan/06/china-sued-piracy-green-dam.

Brenner, Bill. (2005). Myfip's Titan Rain Connection. Retrieved on October 2, 2014, from http://searchsecurity.techtarget.com/news/article/0,289142,sid14_gci1120855,00.html.

Bristow, Michael. (2008). China's Internet "Spin Doctors." Retrieved on September 23, 2014, from http://news.bbc.co.uk/2/hi/asia-pacific/7783640.stm.

Brodkin, Jon. (2007). Government-sponsored cyberattacks on the rise, McAfee says. Retrieved on November 13, 2014, from http://www.networkworld.com/article/2289197/lan-wan/government-sponsored-cyberattacks-on-the-rise--mcafee-says.html.

Bronk, Christopher. (2011). Blown to Bits: China's War in Cyberspace, August–September 2020. Retrieved on February 21, 2015, from http://www.au.af.mil/au/ssq/2011/spring/bronk.pdf.

Broomfield, Emma V. (2003). "Perceptions of Danger: The China Threat Theory." Journal of Contemporary China, Volume 12, Issue 35, 265–84.

Carfano, James. (2008). Combating Enemies Online: State-Sponsored and Terrorist Use of the Internet. Retrieved on September 22, 2014, from http://www.heritage.org/Research/nationalSecurity/upload/bg_2105.pdf.

CBS 60 Minutes. (2009). Sabotaging the System. Retrieved on November 13, 2014, from http://www.youtube.com/watch?v=TkeoxyiBkSM.

Chang, Amy. (2015). China's Maodun: A Free Internet Caged by the Chinese Communist Party. Retrieved on September 21, 2015, from http://www.jamestown.org/programs/chinabrief/single/?tx_ttnews%5Btt_news%5D=43797&cHash=07284c58bf9e001fab71f36808397c29#.Vg_dTHYw9H0.

Channer, Hayley. (2014). Steadying the US rebalance to Asia: The role of Australia, Japan and South Korea. Retrieved on November 21, 2014, from https://www.aspi.org.au/publications/steadying-the-us-rebalance-to-asia-the-role-of-australia,-japan-and-south-korea/SI77_US_rebalance.pdf.

Chase, Michael S.; Engstrom, Jeffrey; Cheung, Tai Ming; Gunness, Kristen A.; Harold, Scott Warren; Puska, Susan; and Berkowitz, Samuel K. (2015). China's Incomplete Military Transformation: Assessing the Weaknesses of the People's Liberation Army (PLA). Retrieved on February 16, 2015, from http://origin.www.uscc.gov/sites/default/files/Research/China%27s%20Incomplete%20Military%20Transformation_2.11.15.pdf.

Cheung, Jennifer. (2015). China's "great firewall" just got taller. Retrieved on September 21, 2015, from https://www.opendemocracy.net/digitaliberties/jennifer-cheung/china's-'great-firewall'-just-got-taller.

Chien, Eric, and O'Gorman, Gavin. (2011). The Nitro Attacks: Stealing Secrets from the Chemical Industry. Retrieved on October 12, 2014, from http://www.symantec.com/content/en/us/enterprise/media/security_response/whitepapers/the_nitro_attacks.pdf.

China blamed after ASIO blueprints stolen in major cyber attack on Canberra HQ. (2013). Retrieved on October 5, 2014, from http://www.abc.net.au/news/2013–05–27/asio-blueprints-stolen-in-major-hacking-operation/4715960.

China calls US culprit in global "Internet war." (2011). Retrieved on October 5, 2014, from http://news.yahoo.com/china-calls-us-culprit-global-internet-war-090540771.html

China cyber-gangs use "vast underground network." (2014). Retrieved on October 3, 2014, from http://www.bbc.com/news/technology-26432616.

China hails UN's anti-Internet terrorism resolution. (2013). Retrieved on October 5, 2014, from http://news.xinhuanet.com/english/china/2013–12/18/c_132978600.htm.

China has "mountains of data" about U.S. cyber attacks: official. (2013). Retrieved on October 5, 2014, from http://www.reuters.com/article/2013/06/05/us-china-usa-hacking-idUSBRE95404L20130605.

China launching "severe" cyber attacks on Taiwan: minister. (2014). Retrieved on October 5, 2014, from http://www.spacedaily.com/reports/China_launching_severe_cyber_attacks_on_Taiwan_minister_999.html.

China orders media giant Sina to "improve censorship." (2015). Retrieved on September 21, 2015, from http://phys.org/news/2015–04-china-media-giant-sina-censorship.html.

China says hit by 500,000 cyberattacks in 2010. (2011). Retrieved on October 5, 2014, from http://www.spacedaily.com/reports/China_says_hit_by_500000_cyberattacks_in_2010_999.html.

China says U.S. routinely hacks Defense Ministry websites. (2013). Retrieved on October 5, 2014, from http://www.reuters.com/article/2013/02/28/us-china-usa-cyber-idUSBRE91R0C120130228.

China Tightens Vice On Internet. (2006). Retrieved on September 22, 2014, from http://cryptome.cn.com/china-vise.htm.

China under suspicion in U.S. for Lockheed hacking. (2011). Retrieved on October 5, 2014, from http://www.reuters.com/article/2011/06/02/us-lockheed-china-idUS-TRE7517B120110602.

China's cyber security under severe threat: report. (2013). Retrieved on October 5, 2014, from http://news.xinhuanet.com/english/china/2013–03/19/c_132246098.htm.

China's Military Strategy. (2015). The State Council Information Office of the People's Republic of China. Retrieved on September 21, 2015, from http://eng.mod.gov.cn/Database/WhitePapers/.

China's New Small Leading Group on Cybersecurity and Internet Management. (2014). Retrieved on October 5, 2014, from http://www.forbes.com/sites/adamsegal/2014/02/27/chinas-new-small-leading-group-on-cybersecurity-and-internet-management/.

China's Proliferation Practices, and the Development of Its Cyber and Space Warfare Capabilities. (2008). Retrieved on September 23, 2014, from http://origin.www.uscc.gov/sites/default/files/transcripts/5.20.08HearingTranscript.pdf.

China's Propaganda and Influence Operations, Its Intelligence Activities That Target the United States, and the Resulting Impacts on U.S. National Security. (2009). Retrieved on September 20, 2014, from http://origin.www.uscc.gov/sites/default/files/transcripts/4.30.09HearingTranscript.pdf.

China's Twelfth Five Year Plan 2011–2015. (2011). Retrieved on October 4, 2014, from http://www.britishchamber.cn/content/chinas-twelfth-five-year-plan-2011–2015-full-english-version.

Chinese hackers attack 745 Vietnam websites in a week: report. (2014). Retrieved on October 18, 2014, from http://www.vir.com.vn/chinese-hackers-attack-745-vietnam-websites-in-a-week-report.html.

Chinese websites to "spread positive energy." (2013). Retrieved on September 21, 2015, from http://news.xinhuanet.com/english/china/2013–10/30/c_132844968.htm.

Chua, Amy. (2004). *World on Fire: How Exporting Free Market Democracy Breeds Ethnic Hatred and Global Instability*. New York: Anchor Books.

Clarke, Richard A., and Knake, Robert K. (2010). *Cyber War*. New York: HarperCollins Publishers.

Clausewitz, Carl von. (2013). On War. Retrieved on April 26, 2015, from http://www.gutenberg.org/files/1946/1946-h/1946-h.htm.

Clayton, Mark. (2012). Stealing US business secrets: Experts ID two huge cyber "gangs" in China. Retrieved on October 15, 2014, from http://www.csmonitor.com/USA/2012/0914/Stealing-US-business-secrets-Experts-ID-two-huge-cyber-gangs-in-China.

Clendenin, Mike. (2010). China Gets A Peek At Microsoft Source Code. Retrieved on November 23, 2014, from http://www.informationweek.com/software/operating-systems/china-gets-a-peek-at-microsoft-source-code/d/d-id/1089702?.

Cliff, Roger. (2011). Anti-Access Measures in Chinese Defense Strategy. Retrieved on December 1, 2014, from http://www.rand.org/content/dam/rand/pubs/testimonies/2011/RAND_CT354.pdf.

Clifford, John. (2014). Conceptual Basis of Future EW. Retrieved on March 21, 2015, from http://tangentlink.com/wp-content/uploads/2014/03/17.-The-Conceptual-Basis-for-Future-EW-John-Clifford.pdf.

Clinton, Hillary. (2011). America's Pacific Century. Retrieved on November 21, 2014, from http://www.foreignpolicy.com/articles/2011/10/11/americas_pacific_century.

CNCERT/CC Annual Report. (2003). Retrieved on November 18, 2014, from http://www.cert.org.cn/publish/english/upload/File/2003CNCERTCCAnnualReport.pdf.

———. (2004). Retrieved on November 18, 2014, from http://www.cert.org.cn/publish/english/upload/File/2004CNCERTCCAnnualReport.pdf.

———. (2005). Retrieved on November 18, 2014, from http://www.cert.org.cn/publish/english/upload/File/2005CNCERTCCAnnualReport.pdf.

———. (2006). Retrieved on November 18, 2014, from http://www.cert.org.cn/publish/english/upload/File/2006AnnualReportByCNCERT.pdf.

———. (2007). Retrieved on November 18, 2014, from http://www.cert.org.cn/publish/english/upload/File/CNCERTAnnualReport2007.pdf.

———. (2008). Retrieved on November 18, 2014, from http://www.cert.org.cn/publish/english/upload/File/CNCERTAnnualReport2008.pdf.

———. (2010). Retrieved on November 18, 2014, from http://www.cert.org.cn/User-Files/File/CNCERTAnnualReport2010v2.pdf.

———. (2012). Retrieved on November 18, 2014, from http://www.cert.org.cn/publish/english/upload/File/APCERT_Annual2012_CNCERT.pdf.

———. (2013). Retrieved on November 18, 2014, from http://www.cert.org.cn/publish/english/upload/File/CNCERT_Annual_Report_2013.pdf.

Collins, Elizabeth Fuller. (2002). Indonesia: A Violent Culture? Retrieved on April 20, 2015, from http://www.ohio.edu/cas/classics/faculty/upload/Indonesia-A-Violent-Culture.pdf.

Committee on Armed Services United States Senate. (2014). Inquiry Into Cyber Intrusions Affecting U.S. Transportation Command Contractors. Retrieved on October 11, 2014, from http://www.armed-services.senate.gov/imo/media/doc/SASC_Cyber report_091714.pdf.

Committee on Foreign Affairs. (2011). Communist Chinese Cyber-Attacks, Cyber-Espionage and Theft of American Technology. Retrieved on October 2, 2014, from http://cryptome.org/2012/10/cn-cyberspy-2011.pdf.

Congressional Research Service Report. (2015). Cyber Intrusion into U.S. Office of Personnel Management: In Brief. Retrieved on October 30, 2016, from http://www.fas.org/sgp/crs/natsec/R44111.pdf.

Constantin, Lucian. (2012). Sykipot Trojan Hijacks Department of Defense authentication smart cards. Retrieved on October 12, 2014, from http://www.infoworld.com/article/2618493/government/sykipot-trojan-hijacks-department-of-defense-authentication-smart-cards.html.

Conti, Gregory, and Surdu, John. (2009). Army, Navy, Air Force, and Cyber—Is it Time for a Cyberwarfare Branch of Military? Retrieved on November 25, 2014, from http://www.rumint.org/gregconti/publications/2009_IAN_12–1_conti-surdu.pdf.

Cost of "Code Red" Rising. (2001). Retrieved on October 2, 2014, from http://edition.cnn.com/2001/TECH/internet/08/08/code.red.II/index.html.

Cox, Matt. (2014). USSOCOM Wants Computer-Draining Tech. Retrieved on December 23, 2014, from http://defensetech.org/2014/11/19/ussocom-wants-computer-draining-tech/#more-24122.

Creemers, Rogier. (2014). China's Internet Gambit. Retrieved on September 21, 2015, from http://www.chinausfocus.com/peace-security/chinas-internet-gambit/.

———. (2015). Ideology Matters: Parsing Recent Changes in China's Intellectual Landscape. Retrieved on February 8, 2015, from https://sinocism.com/?p=11410.

CrowdStrike Intelligence Report. (2014). Putter Panda. Retrieved on October 31, 2014, from http://cdn0.vox-cdn.com/assets/4589853/crowdstrike-intelligence-report-putter-panda.original.pdf.

Cuban, Brian. (2008). Confessions of a Banned Digger. Retrieved on September 23, 2014, from http://www.briancuban.com/confessions-of-a-banned-digger/.

Cyberspace Administration of China. (2015). About Cyberspace of Administration of China. Retrieved on September 21, 2015, from http://www.cac.gov.cn/english/.

Cyberspace Administration of China launches official website. (2015). Retrieved on September 21, 2015, from http://news.xinhuanet.com/english/china/2014–12/31/c_133890303.htm.

Cyberspace Operations. (2013). Joint Publication 3–12. Retrieved on September 21, 2015, from http://www.dtic.mil/doctrine/new_pubs/jp3_12R.pdf.

DCSINT. (2005). Cyber Operations and Cyber Terrorism. Retrieved on November 20, 2014, from http://handle.dtic.mil/100.2/ADA439217.

Delio, Michelle. (2001). Code Blue Targets China Firm. Retrieved on October 25, 2014, from http://archive.wired.com/science/discoveries/news/2001/09/46624.

Dellios, Rosita, and Ferguson, James R. (2013). *China's Quest for Global Order: From Peaceful Rise to Harmonious World*. Lanham, MD: Lexington Books.

Demick, Barbara. (2013). China hacker's angst opens a window onto cyber-espionage. Retrieved on October 13, 2014, from http://articles.latimes.com/2013/mar/12/world/la-fg-china-hacking-20130313.

Denlinger, Paul. (2010). Why China's Web Copycats Succeed. Retrieved on April 20, 2015, from http://www.forbes.com/sites/china/2010/08/19/why-chinas-web-copycats-succeed/.

Department of Defense Dictionary of Military and Associated Terms. (2010). Joint Publication 1–02. Retrieved on September 21, 2015, from http://www.dtic.mil/doctrine/new_pubs/jp1_02.pdf.

Deptula, David A. (2011). China's Active Defense Strategy and its Regional Impact. Retrieved on September 17, 2014, from http://www.uscc.gov/sites/default/files/1.27.11Deptula.pdf.

Desktop Operating System Market Share. (2014). Retrieved on November 23, 2014, from http://www.netmarketshare.com/operating-system-market-share.aspx?qprid=8&qpcustomd=0.

Digital Spying Burdens German-Chinese Relations. (2013). Retrieved on October 13, 2014, from http://www.spiegel.de/international/world/digital-spying-burdens-german-relations-with-beijing-a-885444.html.

Dilanian, Ken. (2012). US Spy Agencies to Detail Cyber Attacks from Abroad. Retrieved on October 13, 2014, from http://articles.latimes.com/2012/dec/06/nation/la-na-cyber-intel-20121207.

Doherty, Stephen; Gegeny, Jozsef; Spasojevic, Branko; and Baltazar, Jonell. (2013). Hidden Lynx—Professional Hackers for Hire. Retrieved on October 25, 2014, from http://www.symantec.com/content/en/us/enterprise/media/security_response/whitepapers/hidden_lynx.pdf.

Dotson, John. (2011). The Confucian Revival in the Propaganda Narratives of the Chinese Government. Retrieved on September 15, 2014, from http://www.uscc.gov/sites/default/files/Research/Confucian_Revival_Paper.pdf.

Dutcher, Dan. (2013). Sykipot Now Targeting US Civil Aviation Sector Information. Retrieved on October 12, 2014, from http://blog.trendmicro.com/trendlabs-security-intelligence/sykipot-now-targeting-us-civil-aviation-sector-information/.

Dwyer, Devin. (2013). "The Nintendo Medal"? New Military Award for Drone Pilots Draws Hill Protest. Retrieved on October 5, 2014, from http://news.yahoo.com/nintendo-medal-military-award-drone-111006056.html.

EADS, ThyssenKrupp attacked by Chinese hackers: report. (2013). Retrieved on October 13, 2014, from http://www.reuters.com/article/2013/02/24/net-us-eads-thyssenkrupp-hacking-idUSBRE91N07M20130224.

East Asia Security Symposium and Conference. (2013). *Maritime Security in the Context of East Asian Strategic Games*. Beijing, China: China Foreign Affairs University.

Echevarria, Antulio J. II. (2002). Clausewitz's Center of Gravity: Changing Our Warfighting Doctrine—Again! Retrieved on April 26, 2015, from http://www.strategicstudiesinstitute.army.mil/Pubs/display.cfm?pubID=363.

Edwards, Sean J. A. (2004). Swarming and the Future of Warfare. Retrieved on February 25, 2015, from http://www.rand.org/content/dam/rand/pubs/rgs_dissertations/2005/RAND_RGSD189.pdf.

Electronic Warfare. (1999). Air Force Operational Doctrine. Retrieved on March 21, 2015, from http://www.globalsecurity.org/military/library/policy/usaf/afdd/2–5-1/afdd2–5-1-draft.pdf.

———. (2007). Joint Publication 3–13.1. Retrieved on December 23, 2014, from http://www.fas.org/irp/doddir/dod/jp3–13–1.pdf.

Electronic Warfare in Operations. (2009). Field Manual No. 3–36. Retrieved on December 23, 2014, from http://usacac.army.mil/cac2/Repository/FM336/FM336.pdf.

Elgin, Ben, and Epstein, Keith. (2008). Tech—Network security breaches & NASA. Retrieved on October 5, 2014, from http://spoonfeedin.blogspot.com.au/2008/11/tech-network-security-breaches-nasa.html.

Espiner, Tom. (2005). Security Experts Lift Lid On Chinese Hack Attacks. Retrieved on October 2, 2014, from http://news.cnet.com/Security-experts-lift-lid-on-Chinese-hack-attacks/2100–7349_3–5969516.html.

Esposito, Richard. (2012). "Astonishing" Cyber Espionage Threat from Foreign Governments: British Spy Chief. Retrieved on October 2, 2014, from http://news.yahoo.com/astonishing-cyber-espionage-threat-foreign-governments-british-spy-191653463--abc-news-topstories.html.

Evers, Marco. (2005). The Weapon of Sound: Sonic Canon Gives Pirates an Earful. Retrieved on February 25, 2015, from http://www.spiegel.de/international/spiegel/the-weapon-of-sound-sonic-canon-gives-pirates-an-earful-a-385048.html.

Faiola, Anthony. (2005). Anti-Japanese Hostilities Move to the Internet. Retrieved on September 23, 2014, from http://www.washingtonpost.com/wp-dyn/content/article/2005/05/09/AR2005050901119.html.

FBI warns industry of Chinese cyber campaign. (2014). Retrieved on October 17, 2014, from http://www.washingtonpost.com/world/national-security/fbi-warns-industry-of-chinese-cyber-campaign/2014/10/15/0349a00a-54b0–11e4-ba4b-f6333e2c0453_story.html.

Feakin, Tobias. (2013). Enter the Cyber Dragon: Understanding Chinese intelligence agencies' cyber capabilities. Retrieved on April 20, 2015, from https://www.aspi.org.au/publications/special-report-enter-the-cyber-dragon-understanding-chinese-intelligence-agencies-cyber-capabilities/10_42_31_AM_SR50_chinese_cyber.pdf.

Flaherty, Anne. (2013). US Swipes at China for Hacking Allegations. Retrieved on October 5, 2014, from http://news.yahoo.com/us-swipes-china-hacking-allegations-193407762.html.

Flaherty, Mary Pat, Samenow, Jason, and Rein, Lisa. (2014). Chinese hack U.S. weather systems, satellite network. Retrieved on December 2, 2014, from http://www.washingtonpost.com/local/chinese-hack-us-weather-systems-satellite-network/2014/11/12/bef1206a-68e9–11e4-b053–65cea7903f2e_story.html.

Fletcher, Owen. (2009). China Warns About Return of Destructive Panda Virus. Retrieved on December 5, 2014, from http://www.pcworld.com/article/183272/article.html.

Fox, Stuart. (2009). July 4th Hacker Attack Targeted Major U.S. Government Sites. Retrieved on October 9, 2014, from http://www.popsci.com.au/scitech/article/2009–07/hacker-attack-slows-down-holiday-web-surfing.

Franciska, Christine. (2014). New voting power of Chinese Indonesians. Retrieved on April 20, 2015, from http://www.bbc.com/news/world-asia-27991754.

Fritz, Jason. (2008). How China Will Use Cyber Warfare to Leapfrog in Military Competitiveness. Retrieved on September 20, 2014, from http://epublications.bond.edu.au/cgi/viewcontent.cgi?article=1110&context=cm.

———. (2009). Hacking Nuclear Command and Control. Retrieved on September 20, 2014, from http://icnnd.org/Documents/Jason_Fritz_Hacking_NC2.pdf.

———. (2013a). Satellite Hacking: A Guide for the Perplexed. Retrieved on September 20, 2014, from http://epublications.bond.edu.au/cgi/viewcontent.cgi?article=1131&context=cm.

———. (2013b). The Semantics of Cyber Warfare. Retrieved on September 20, 2014, from http://epublications.bond.edu.au/cgi/viewcontent.cgi?article=1041&context=eassc_publications.

Gady, Franz-Stefan. (2015). Why the PLA Revealed Its Secret Plans for Cyber War. Retrieved on April 20, 2015, from http://thediplomat.com/2015/03/why-the-pla-revealed-its-secret-plans-for-cyber-war/.

GAO Critical Infrastructure Protection Commercial Satellite Security Should Be More Fully Addressed. (2002). Retrieved on February 12, 2013, from http://www.gao.gov/assets/240/235485.pdf.

Gardner, David. (2008). Satellite Pictures Reveal Massive Chinese Nuclear Submarine Base, Says Pentagon. Retrieved on September 23, 2014, from http://www.dailymail.co.uk/news/article-563405/Satellite-pictures-reveal-massive-Chinese-nuclear-submarine-base-says-Pentagon.html.

Gertz, Bill. (2000). Hackers linked to China stole Los Alamos documents. Retrieved on November 23, 2014, from http://seclists.org/isn/2000/Aug/14.

———. (2014). Top Gun takeover: Stolen F-35 secrets showing up in China's stealth fighter. Retrieved on October 22, 2014, from http://www.washingtontimes.com/news/2014/mar/13/f-35-secrets-now-showing-chinas-stealth-fighter/?page=all.

Glanz, James, and Markoff, John. (2010). Vast Hacking by a China Fearful of the Web. Retrieved on October 15, 2014, from http://www.nytimes.com/2010/12/05/world/asia/05wikileaks-china.html?pagewanted=all&_r=0.

Gold, Michael. (2013). Taiwan a "testing ground" for Chinese cyber army. Retrieved on April 20, 2015, http://www.reuters.com/article/2013/07/19/net-us-taiwan-cyber-idUSBRE96H1C120130719.

Gold, Michael, and Wu, J. R. (2015). Taiwan seeks stronger cyber security ties with US to counter China threat. Retrieved on April 20, 2015, from http://www.reuters.com/article/2015/03/30/us-taiwan-cybersecurity-iduskbn0mq11v20150330.

Goldkorn, Jeremy. (2011). YouTube = Youku? Websites and Their Chinese Equivalents. Retrieved on April 20, 2015, from http://www.fastcompany.com/1715042/youtube-youku-websites-and-their-chinese-equivalents.

Goodin, Dan. (2009). Hackers declare war on international forensics tool: Microsoft's COFEE decaffeinated. Retrieved on December 23, 2014, from http://www.theregister.co.uk/2009/12/14/microsoft_cofee_vs_decaf/.

Gorman, Siobhan. (2009). Electricity Grid in U.S. Penetrated By Spies. Retrieved on October 9, 2014, from http://online.wsj.com/articles/SB123914805204099085.

———. (2012). Chinese Hackers Suspected In Long-Term Nortel Breach. Retrieved on October 2, 2014, from http://online.wsj.com/article/SB10001424052970203363504577187502201577054.html.

Gorman, Siobhan, Dreazen, Yochi J., and Cole, August. (2009). Insurgents Hack US Drones. Retrieved on October 7, 2014, from http://online.wsj.com/article/SB126102247889095011.html.

Government "may have hacked IMF." (2011). Retrieved on October 2, 2014, from http://www.bbc.co.uk/news/technology-13748488.

Gragido, Will. (2012). Lions at the Watering Hole—The "VOHO" Affair. Retrieved on December 5, 2014, from https://blogs.rsa.com/lions-at-the-watering-hole-the-voho-affair/.

Greenemeier, Larry. (2007). China's Cyber Attacks Signal New Battlefield Is Online. Retrieved on December 2, 2014, from http://www.scientificamerican.com/article/chinas-cyber-attacks-sign/.

Group of Governmental Experts. (2013). Developments in the Field of Information and Telecommunications in the Context of International Security. Retrieved on November 22, 2014, from http://www.un.org/ga/search/view_doc.asp?symbol=A/68/98.

Grow, Brian, and Hosenball, Mark. (2011). Special report: In cyberspy vs. cyberspy, China has the edge. Retrieved on October 13, 2014, from http://www.reuters.com/article/2011/04/14/us-china-usa-cyberespionage-idUSTRE73D24220110414.

Guo, Lei; Gu, Caiyu; and Wu, Lan. (2011). Why China established "Online Blue Army." Retrieved on November 23, 2014, from http://english.people.com.cn/90001/90780/7423270.html.

Gutteberg, Odd. (1993). Telektronikk 4.92 Satellite Communications. Retrieved on March 26, 2013, from http://www.telenor.com/wp-content/uploads/2012/05/T92_4.pdf.

Hammond, Rupert J. (2014). Recent Developments in China's Relations with Taiwan and North Korea. Retrieved on April 20, 2014, from http://www.us-taiwan.org/reports/2014_june05_cross-strait_economic_political_issues_testimony_to_uscc.pdf.

Hansen, Simon. (2014). Australia and great power cyber strategy after APEC. Retrieved on September 21, 2015, from http://www.aspistrategist.org.au/great-powers-australia-and-cyber-strategy-after-apec-and-the-g20/.

Harris, Shane. (2008). China's Cyber-Militia. Retrieved on October 22, 2014, from http://www.nationaljournal.com/magazine/china-s-cyber-militia-20080531.

Harrison, Linda. (2000). Bedroom NASA hacker set to bite pillow in choky. Retrieved on November 21, 2014, from http://www.theregister.co.uk/2000/09/22/bedroom_nasa_hacker_set/.

HBGary. (2010). Operation Aurora. Retrieved on October 11, 2014, from http://www.hbgary.com/sites/default/files/publications/WhitePaper%20HBGary%20Threat%20Report,%20Operation%20Aurora.pdf.

———. (2011). China's State-sponsored Espionage. Retrieved on April 20, 2015, from http://cryptome.org/0003/hbg/HGB-CN-Spy.zip.

Heickerö, Roland. (2012). *The Dark Sides Of The Internet: On Cyber Threats and Information Warfare*. Frankfurt am Main, Germany: Peter Lang International Academic Publishers.

Henderson, Scott. (2007a). The Dark Visitor: Inside the World of Chinese Hackers. Northwestern University, US: Lulu.com.

———. (2007b). Javaphile, Buddhism, and . . . The Public Security Bureau? Retrieved on October 13, 2014, from http://www.thedarkvisitor.com/2007/12/javaphile-buddhism-andthe-public-security-bureau/.

———. (2009). Panda Burning Incense author to work in "Computer Security." Retrieved on December 5, 2014, from http://www.thedarkvisitor.com/2009/07/panda-burning-incense-author-to-work-in-computer-security/.

———. (2011). US #1 perp attacking China's classified networks. Retrieved on November 1, 2014, from http://www.thedarkvisitor.com/2011/03/us-1-perp-attacking-chinas-classifed-networks/.

Hoffman, Mike. (2013). Hagel Calls Out China on Cyber Attacks. Retrieved on October 5, 2014, from http://defensetech.org/2013/06/03/hagel-calls-out-china-on-cyber-attacks/#idc-container.

———. (2014). Cyber is Likely Winner of 2015 Budget. Retrieved on October 5, 2014, from http://defensetech.org/2014/02/24/cyber-is-likely-winner-of-2015-budget/#more-22396.

Honker Union of China to launch network attacks against Japan is a rumor. (2010). Retrieved on October 13, 2014, from http://www.chinahush.com/2010/09/15/honker-union-of-china-to-launch-network-attack-against-japan-is-a-rumor/.

Hopkins, Nick. (2011). UK developing cyber-weapons programme to counter cyber war threat. Retrieved on November 13, 2014, from http://www.theguardian.com/uk/2011/may/30/military-cyberwar-offensive.

———. (2012). US and China engage in cyber war games. Retrieved on October 5, 2014, from http://www.theguardian.com/technology/2012/apr/16/us-china-cyber-war-games.

Hosenball, Mark, and Eckert, Paul. (2011). China under suspicion in U.S. for Lockheed hacking. Retrieved on October 2, 2014, from http://www.reuters.com/article/2011/06/02/us-lockheed-china-idUSTRE7517B120110602.

Hsiao, Russell. (2013). Critical Node: Taiwan's Cyber Defense and Chinese Cyber-Espionage. Retrieved on October 2, 2014, from http://www.jamestown.org/single/?tx_ttnews%5Btt_news%5D=41721&no_cache=1#.VC1EsWwcRH0.

Hui, Li, and Wee, Sui-Lee. (2014). Chinese militants get Islamic State "terrorist training"—media. Retrieved on September 22, 2014, from http://uk.reuters.com/article/2014/09/22/uk-china-xinjiang-idUKKCN0HH10120140922.

Hypponen, Mikko. (2006). Malware Goes Mobile. Retrieved on December 6, 2014, from http://www.cs.virginia.edu/~robins/Malware_Goes_Mobile.pdf.

IDC. (2014). Worldwide Smartphone Shipments Top One Billion Units for the First Time, According to IDC. Retrieved on December 8, 2014, from http://www.idc.com/getdoc.jsp?containerId=prUS24645514.

Information Office of the State Council of the People's Republic of China. (2000). China's National Defense in 2000 (White Paper). Retrieved on January 17, 2015, from http://www.china.org.cn/e-white/2000/index.htm.

———. (2002). China's National Defense in 2002 (White Paper). Retrieved on January 17, 2015, from http://www.china.org.cn/e-white/20021209/index.htm.

———. (2004). China's National Defense in 2004 (White Paper). Retrieved on January 17, 2015, from http://www.china.org.cn/e-white/20041227/index.htm.

———. (2006). China's National Defense in 2006 (White Paper). Retrieved on January 17, 2015, from http://www.china.org.cn/english/features/book/194421.htm.

———. (2008). China's National Defense in 2008 (White Paper). Retrieved on January 17, 2015, from http://www.china.org.cn/government/whitepaper/node_7060059.htm.

————. (2010) . China's National Defense in 2010 (White Paper). Retrieved on January 17, 2015, from http://www.china.org.cn/government/whitepaper/node_7114675.htm.

————. (2013). The Diversified Employment of China's Armed Forces (White Paper). Retrieved on October 7, 2014, from http://www.china.org.cn/government/white-paper/node_7181425.htm.

Information Operations. (2012). Joint Publication 3–13. Retrieved on September 21, 2015, from http://www.dtic.mil/doctrine/new_pubs/jp3_13.pdf.

Information Operations Roadmap. (2003). Declassified U.S. Government document. Retrieved on September 16, 2014, from http://www.gwu.edu/~nsarchiv/NSAEBB/NSAEBB177/info_ops_roadmap.pdf.

Information Warfare Monitor. (2009). Tracking GhostNet: Investigating a Cyber Espionage Network. Retrieved on October 7, 2014, from http://www.scribd.com/doc/13731776/Tracking-GhostNet-Investigating-a-Cyber-Espionage-Network.

————. (2010). Shadows in the Cloud: Investigating Cyber Espionage 2.0. Retrieved on October 7, 2014, from http://www.nartv.org/mirror/shadows-in-the-cloud.pdf.

Ingersoll, Geoffrey. (2013). NSA Says It Foiled Plot To Destroy Our Economy By Bricking Computers Across The US. Retrieved on October 5, 2014, from http://www.businessinsider.com.au/nsa-says-foiled-china-cyber-plot-2013–12.

Interim Staff Report. (2016). The Science, Space, and Technology Committee's Investigation of FDIC's Cybersecurity. Retrieved on October 30, 2016, from https://science.house.gov/sites/republicans.science.house.gov/files/documents/Final%20GOP%20Interim%20Staff%20Report%207–12–16.pdf.

International Institute for Strategic Studies. (2015). *The Military Balance 2015*. London: Routledge.

Jing, de Jong-Chen. (2014). U.S.-China Cybersecurity Relations: Understanding China's Current Environment. Retrieved on February 8, 2015, from http://journal.georgetown.edu/u-s-china-cybersecurity-relations-understanding-chinas-current-environment/.

Joint Electromagnetic Spectrum Management Operations. (2012). Joint Publication 6–01. Retrieved on September 21, 2015, from http://www.dtic.mil/doctrine/new_pubs/jp6_01.pdf.

Jowitt, Tom. (2013). Eurofighter Maker EADS Attacked—Chinese Hackers Blamed. Retrieved on October 14, 2014, from http://www.techweekeurope.co.uk/news/chinese-hackers-eads-eurofighte-108651.

Kaspersky Lab. (2013a). The NetTraveler. Retrieved on October 26, 2014, from http://25zbkz3k00wn2tp5092n6di7b5k.wpengine.netdna-cdn.com/files/2014/07/kaspersky-the-net-traveler-part1-final.pdf.

————. (2013b). The Icefog APT: A Tale of Cloak and Three Daggers. Retrieved on October 26, 2014, from http://kasperskycontenthub.com/wp-content/uploads/sites/43/vlpdfs/icefog.pdf.

Keck, Zachary. (2013). Forget Japan: China's ADIZ Threatens Taiwan. Retrieved on April 18, 2015, from http://thediplomat.com/2013/12/forget-japan-chinas-adiz-threatens-taiwan/.

Kelly, Tim, and Lee, Melanie. (2012). Sony PlayStation certificate sparks talk China may lift console ban. Retrieved on October 5, 2014, from http://www.reuters.com/article/2012/11/07/us-sony-ps3-china-idUSBRE8A60RH20121107.

King, Gary; Pan, Jennifer; and Roberts, Margaret E. (2014). Reverse-engineering censorship in China: Randomized experimentation and participant observation. Retrieved on September 15, 2014, from http://cryptome.org/2014/08/reverse-eng-cn-censorship.pdf.

Klotz, Irene. (2013). NASA steps up security after arrest of former contractor. Retrieved on October 5, 2014, from http://www.reuters.com/article/2013/03/20/us-space-espionage-idUSBRE92J1FR20130320.

Knox, Olivier. (2013). Drones have killed 4,700, U.S. senator says. Retrieved on February 26, 2013, from http://news.yahoo.com/blogs/ticket/drones-killed-4–700-u-senator-says-141143752--politics.html.

Krekel, Bryan. (2009). Capability of the People's Republic of China to Conduct Cyber Warfare and Computer Network Exploitation. Retrieved on September 30, 2014, from http://www2.gwu.edu/~nsarchiv/NSAEBB/NSAEBB424/docs/Cyber-030.pdf.

Krekel, Bryan; Adams, Patton; and Bakos, George. (2012). Occupying the Information High Ground: Chinese Capabilities for Computer Network Operations and Cyber Espionage. Retrieved on September 22, 2014, from http://origin.www.uscc.gov/sites/default/files/Research/USCC_Report_Chinese_Capabilities_for_Computer_Network_Operations_and_Cyber_%20Espionage.pdf.

Krepinevich, Andrew F. (2010). Why AirSea Battle? Retrieved on October 7, 2014, from http://www.csbaonline.org/publications/2010/02/why-airsea-battle/.

Lam, Lana. (2013a). Exclusive: NSA targeted China's Tsinghua University in extensive hacking attacks, says Snowden. Retrieved on October 5, 2014, from http://www.scmp.com/news/china/article/1266892/exclusive-nsa-targeted-chinas-tsinghua-university-extensive-hacking.

———. (2013b). Exclusive: US hacked Pacnet, Asia Pacific fibre-optic network operator, in 2009. Retrieved on October 5, 2014, from http://www.scmp.com/news/hong-kong/article/1266875/exclusive-us-hacked-pacnet-asia-pacific-fibre-optic-network-operator.

Lange, Jason, and Volz, Dustin. (2016). Likely hack of U.S. banking regulator by China covered up: probe. Retrieved on October 30, 2016, from http://www.reuters.com/article/us-cyber-fdic-china-idUSKCN0ZT20M.

Laurie, Adam. (2009). Satellite Hacking for Fun and Profit. Retrieved on February 14, 2013, from http://www.securitytube.net/video/263.

Lawrence, Dune, and Riley, Michael. (2013). A Chinese Hacker's Identity Unmasked. Retrieved on October 11, 2014, from http://www.businessweek.com/articles/2013–02–14/a-chinese-hackers-identity-unmasked.

Lawrence, Susan V., and Martin, Michael F. (2013). Understanding China's Political System. Retrieved on September 24, 2014, from http://fas.org/sgp/crs/row/R41007.pdf.

Le Mière, Christian. (2013). China's Unarmed Arms Race: Beijing's Maritime Build-Up Isn't What It Appears. Retrieved on April 7, 2015, from http://www.foreignaffairs.com/articles/139609/christian-le-miere/chinas-unarmed-arms-race.

Lee, Dave. (2013). The Comment Group: The hackers hunting for clues about you. Retrieved on October 15, 2014, from http://www.bbc.com/news/business-21371608.

Lee, Eloise, and Johnson, Robert. (2012). China's J-20 and the American F-22 Raptor—You Are Not Seeing Double. Retrieved on October 22, 2014, from http://www.businessinsider.com.au/the-similarities-between-the-j-20-heads-up-display-and-that-on-the-f-22-are-striking-2012–6.

Leftly, Mark. (2014). G20 summit: Enter Putin. Retrieved on November 21, 2014, from http://www.independent.co.uk/news/world/politics/g20-summit-enter-putin-accompanied-by-four-warships-to-the-sound-of-mockery-9862465.html.

Lemon, Sumner. (2007). Report: Chinese police arrest eight for computer virus. Retrieved on December 5, 2014, from http://www.infoworld.com/article/2659010/security/report--chinese-police-arrest-eight-for-computer-virus.html.

Lemos, Robert. (2011). Espionage network exploiting Adobe Reader flaw. Retrieved on October 12, 2014, from http://www.infoworld.com/article/2618792/malware/espionage-network-exploiting-adobe-reader-flaw.html.

Levin, Carl, and McCain, John. (2012). Senate Armed Services Committee Releases Report on Counterfeit Electronic Parts. Retrieved on October 2, 2014, from http://www.scribd.com/doc/94426658/SASC-Counterfeit-Electronics-Report-05–21–12.

Lewis, James Andrew. (2010). The Electrical Grid as a Target for Cyber Attack. Retrieved on December 2, 2014, from http://csis.org/files/publication/100322_ElectricalGridAsATargetforCyberAttack.pdf.

Lewis, Leo. (2011). China's Blue Army of 30 computer experts could deploy cyber warfare on foreign powers. Retrieved on October 5, 2014, from http://www.theaustralian.com.au/technology/chinas-blue-army-could-conduct-cyber-warfare-on-foreign-powers/story-e6frgakx-1226064132826?nk=8b70171fe663 555aa926d533210b3c43.

Lewis, Peter. (1994). Computer Snoopers Imperil Pentagon Files, Experts Say. Retrieved on December 5, 2014, from http://query.nytimes.com/gst/fullpage.html?res=9F04E3DD143EF932A15754C0A962958260.

Leyden, John. (2007). France blames China for hack attacks. Retrieved on October 2, 2014, from http://www.theregister.co.uk/2007/09/12/french_cyberattacks/.

Li, Heng. (2003). Microsoft Gives Chinese Government Access to Windows Source Code. Retrieved on November 23, 2014, from http://english.peopledaily.com.cn/200303/04/eng20030304_112657.shtml.

Li, Yancheng. (2011). Hacker attacks on gov't websites increasing. Retrieved on November 1, 2014, from http://english.peopledaily.com.cn/90001/98649/7319285.html.

Liang, Qiao, and Xiangsui, Wang. (1999). Unrestricted Warfare. Retrieved on September 16, 2014, from http://www.terrorism.com/documents/unrestricted.pdf.

Libicki, Martin C. (1994). The Small and the Many. Retrieved on February 25, 2015, from http://www.rand.org/content/dam/rand/pubs/monograph_reports/MR880/MR880.ch8.pdf.

Lourdeau, Keith. (2004). Virtual Threat, Real Terror: Cyberterrorism in the 21st Century. Retrieved on December 2, 2014, from http://www.globalsecurity.org/security/library/congress/2004_h/040224-lourdeau.htm.

Luard, Tim. (2005). China's Spies Come Out From The Cold. Retrieved on October 2, 2014, from http://news.bbc.co.uk/2/hi/asia-pacific/4704691.stm.

Luckycat Redux: Inside an APT Campaign with Multiple Targets in India and Japan. (2012). Retrieved on October 13, 2014, from http://www.trendmicro.com/cloud-content/us/pdfs/security-intelligence/white-papers/wp_luckycat_redux.pdf.

Lynn, William J. III. (2009). Deputy Secretary of Defense Speech. Retrieved on November 25, 2014, from http://www.defense.gov/speeches/speech.aspx?speechid=1399.

Macri, Giuseppe. (2015). Watch This F-18 Pick Up An Airborne Cruise Missile And Guide It Into A Moving Ship. Retrived on March 9, 2015, from http://news.yahoo.com/watch-f-18-pick-airborne-cruise-missile-guide-170458025.html.

Mandiant. (2013). APT1: Exposing One of China's Cyber Espionage Units. Retrieved on October 12, 2014, from http://intelreport.mandiant.com/Mandiant_APT1_Report.pdf.

Mao, Tse-Tung [Zedong]. (2000). *On Guerrilla Warfare* (trans. Samuel B. Griffith). Chicago: University of Illinois Press.

Mark, David. (2008). Scientists one step closer to invisibility cloak. Retrieved on March 21, 2015, from http://www.abc.net.au/news/stories/2008/08/11/2330897.htm.

Markoff, John. (2009). Old Trick Threatens the Newest Weapons. Retrieved on December 2, 2014, from http://www.nytimes.com/2009/10/27/science/27trojan.html?_r=2&ref=science&pagewanted=all&.

Martin, Paul K. (2011). Inadequate Security Practices Expose Key NASA Network to Cyber Attack. Retrieved on October 2, 2014, from http://oig.nasa.gov/audits/reports/FY11/IG-11–017.pdf.

———. (2012). NASA Cybersecurity: An Examination of the Agency's Information Security. Retrieved on January 26, 2017 from https://oig.nasa.gov/congressional/FINAL_written_statement_for_%20IT_%20hearing_February_26_edit_v2.pdf.

Martin, Rick. (2011). Hundreds of Vietnamese Websites Hacked After Island Dispute With China. Retrieved on October 18, 2014, from https://www.techinasia.com/vietnam-china-hack/.

Martinez, Luis; Meek, James Gordon; Ross, Brian; and Ferran, Lee. (2013). Major U.S. Weapons Compromised By Chinese Hackers, Report Warns. Retrieved on October 5, 2014, from http://news.yahoo.com/major-u-weapons-compromised-chinese-hackers-report-warns-215309601--abc-news-topstories.html.

Maynor, David, and Graham, Robert. (2006). SCADA Security and Terrorism: We're Not Crying Wolf. Retrieved on December 2, 2014, from http://www.blackhat.com/presentations/bh-federal-06/BH-Fed-06-Maynor-Graham-up.pdf.

McAfee. (2009). Virtual Criminology Report 2009. Retrieved on November 13, 2014, from http://iom.invensys.com/EN/pdfLibrary/McAfee/WP_McAfee_Virtual_Criminology_Report_2009_03–10.pdf.

———. (2011). Global Energy Cyberattacks: "Night Dragon." Retrieved on October 2, 2014, from http://www.mcafee.com/au/resources/white-papers/wp-global-energy-cyberattacks-night-dragon.pdf.

———. (2014). Net Losses: Estimating the Global Cost of Cybercrime. Retrieved on April 20, 2015, from http://www.mcafee.com/au/resources/reports/rp-economic-impact-cybercrime2.pdf.

McCoy, Kevin. (2013). Target confirms encrypted PIN data stolen. Retrieved on December 2, 2014, from http://www.usatoday.com/story/money/business/2013/12/27/target-confirms-encrypted-pin-data-stolen/4219415/.

McDonald, Joe, and McGuirk, Rod. (2012). Australia bans Chinese company from Web network. Retrieved on October 2, 2014, from http://news.yahoo.com/australia-bans-chinese-company-network-070505601.html.

McGarry, Brendan. (2013a). China's Cyber Attacks Threaten Social Order: Analyst. Retrieved on October 5, 2014, from http://defensetech.org/2013/05/21/chinas-cyber-attacks-threaten-social-order-analyst/#more-20397.

———. (2013b). Chinese Satellite Grabs Another in Orbit. Retrieved on October 5, 2014, from http://defensetech.org/2013/10/03/chinese-satellite-grabs-another-in-orbit/#more-21449.

Metz, Steven, and Kievit, James. (1995). Strategy and the Revolution in Military Affairs: From Theory to Policy. Retrieved on December 23, 2014, from http://www.au.af.mil/au/awc/awcgate/ssi/stratrma.pdf.

Military Information Support Operations. (2010). Joint Publication 3–13.2. Retrieved on September 21, 2015, from https://www.pksoi.org/document_repository/Lessons/JP3_13_2_MISO_(20-Dec-2011)-LMS-1255.pdf.

Miller, Greg. (2011). Under Obama, an emerging global apparatus for drone killing. Retrieved on November 21, 2014, from http://www.washingtonpost.com/national/national-security/under-obama-an-emerging-global-apparatus-for-drone-killing/2011/12/13/gIQANPdILP_story.html.

Mitnick, Kevin D. (2002). The Art of Deception. Retrieved on December 6, 2015, from http://www.scis.nova.edu/~cannady/ARES/mitnick.pdf.

Monroe, John S. (2009). Cyber Command: So much still to know. Retrieved on November 25, 2014, from http://fcw.com/Articles/2009/07/06/buzz-cyber-command.aspx.

Moran, Ned; Vashisht, Sai Omkar; Scott, Mike; and Haq, Thoufique. (2013). Operation Ephemeral Hydra: IE Zero-Day Linked to DeputyDog Uses Diskless Method. Retrieved on October 31, 2014, from http://www.fireeye.com/blog/technical/cyber-exploits/2013/11/operation-ephemeral-hydra-ie-zero-day-linked-to-deputydog-uses-diskless-method.html.

Moran, Ned, and Villeneuve, Nart. (2013). Operation DeputyDog: Zero-Day (CVE-2013–3893) Attack Against Japanese Targets. Retrieved on October 31, 2014, from http://www.fireeye.com/blog/technical/cyber-exploits/2013/09/operation-deputy-dog-zero-day-cve-2013–3893-attack-against-japanese-targets.html.

Moskvitch, Katia. (2012). Pinterest clones flooding Chinese web space. Retrieved on April 20, 2015, from http://www.bbc.com/news/technology-17812903.

Mulvenon, James C. (1999). The PLA and Information Warfare. Retrieved on January 5, 2015, from http://www.rand.org/content/dam/rand/pubs/conf_proceedings/CF145/CF145.chap9.pdf.

———. (2005). Breaching the Great Firewall. Retrieved on September 22, 2014, from http://www.uscc.gov/sites/default/files/4.14.05mulvenon_james_wrts.pdf.

———. (2013). Chinese Cyber Espionage. Retrieved on December 8, 2014, from http://www.cecc.gov/sites/chinacommission.house.gov/files/CECC%20Hearing%20-%20

Chinese%20Hacking%20-%20James%20Mulvenon%20Written%20Statement.pdf.

Murray, Craig. (2013). Taiwan's Declining Defense Spending Could Jeopardize Military Preparedness. Retrieved on April 20, 2015, from http://www.uscc.gov/sites/default/files/Research/Taiwan%E2%80%99s%20Declining%20Defense%20Spending%20Could%20Jeopardize%20Military%20Preparedness_Staff%20Research%20Backgrounder.pdf.

Nanjing massacre diary author Azuma Shiro died. (2006). Retrieved on April 20, 2015, from http://en.people.cn/200601/04/eng20060104_232798.html.

Navrozov, Lev. (2005). Chinese Geostrategy: The Assassin's Mace. Retrieved on December 2, 2014, from http://www.worldtribune.com/worldtribune/05/sound2453667.5770833334.html.

NCW Roadmap. (2007). Australian Government; Department of Defence. Retrieved on December 23, 2014, from http://www.defence.gov.au/capability/ncwi/docs/2007NCW_roadmap.pdf.

Network-Centric Warfare: Creating a Decisive Warfighting Advantage. (2003). Retrieved on December 23, 2014, from https://www.hsdl.org/?view&did=446193.

New Claims of Hacking as US, China Discuss Cybersecurity. (2014). Retrieved on October 5, 2014, from http://www.voanews.com/content/ny-times-chinese-hackers-searched-data-us-government-workers/1954504.html.

Norris, Pat. (2010). *Watching Earth from Space*. Chichester, UK: Praxis Publishing.

Nortel hit by suspected Chinese cyberattacks for a decade. (2012). Retrieved on October 2, 2014, from http://www.cbc.ca/news/business/story/2012/02/14/nortel-chinese-hackers.html.

Norton-Taylor, Richard. (2007). Titan Rain: How Chinese Hackers Targeted Whitehall. Retrieved on October 2, 2014, from http://www.guardian.co.uk/technology/2007/sep/04/news.internet.

Novetta. (2014). Operation SMN: Axiom Threat Actor Group Report. Retrieved on October 29, 2014, from http://www.novetta.com/files/9714/1446/8199/Executive_Summary-Final_1.pdf.

Nye, Joseph S. Jr. (2004). Soft Power: The Means to Success in World Politics. Retrieved on April 20, 2015, from https://webfiles.uci.edu/schofer/classes/2010soc2/readings/8%20Nye%20Soft%20Power%20Ch%201.pdf.

O'Gorman, Gavin, and McDonald, Geoff. (2012). The Elderwood Project. Retrieved on October 12, 2014, from http://www.symantec.com/content/en/us/enterprise/media/security_response/whitepapers/the-elderwood-project.pdf.

ONCIX, Office of the National Counterintelligence Executive. (2011). Foreign Spies Stealing US Economic Secrets In Cyberspace. Retrieved on October 2, 2014, from http://www.ncix.gov/publications/reports/fecie_all/Foreign_Economic_Collection_2011.pdf.

Onley, Dawn, and Wait, Patience. (2006). Red Storm Rising. Retrieved on October 25, 2014, from http://gcn.com/articles/2006/08/17/red-storm-rising.aspx.

Oshii, Mamoru. (1995). *Ghost in the Shell*. Kokubunji, Tokyo : Production I.G.

Pace, Julie. (2013). Obama says US, China must develop cyber rules. Retrieved on October 5, 2014, from http://news.yahoo.com/obama-says-us-china-must-develop-cyber-rules-034227038.html.

Page, Jeremy. (2011). After Protest Video, U.S. Envoy's Name Censored Online. Retrieved on April 20, 2015, from http://blogs.wsj.com/chinarealtime/2011/02/24/after-protest-video-u-s-envoys-name-censored-online/.

Paget, Francois. (2013). Hacking Summit Names Nations With Cyberwarfare Capabilities. Retrieved on November 13, 2014, from http://blogs.mcafee.com/mcafee-labs/hacking-summit-names-nations-with-cyberwarfare-capabilities.

Pagliery, Jose. (2016). China hacked the FDIC—and US officials covered it up, report says. Retrieved on October 30, 2016, from http://money.cnn.com/2016/07/13/technology/china-fdic-hack/.

Panda, Ankit. (2014). Xi Jinping: China Should Become a "Cyber Power." Retrieved on October 5, 2014, from http://thediplomat.com/2014/03/xi-jinping-china-should-be-come-a-cyber-power/.

Passeri, Paolo. (2012). Philippines and China, on the Edge of a New Cyber Conflict? Retrieved on October 2, 2014, from http://hackmageddon.com/2012/05/01/philip-pines-and-china-on-the-edge-of-a-new-cyber-conflict/.

Pasternack, Alex. (2008). When Nature Won't Cooperate in China, Photoshop! Retrieved on September 23, 2014, from http://www.treehugger.com/files/2008/02/fake_photo_tibet_railway_antelope_greenwashing.php.

Paterson, Tony. (2013). Checking in with "Royal Concierge": GCHQ ran hotel surveillance ring to spy on diplomats and delegations. Retrieved on Aril 20, 2015, from http://www.independent.co.uk/news/world/europe/checking-in-with-royal-con-cierge-gchq-ran-hotel-surveillance-ring-to-spy-on-diplomats-and-delegations-8945520.html.

Perlroth, Nicole. (2012). In Cyberattack on Saudi Firm, U.S. Sees Iran Firing Back. Retrieved on December 2, 2014, from http://www.nytimes.com/2012/10/24/busi-ness/global/cyberattack-on-saudi-oil-firm-disquiets-us.html?adxnnl=1&page-wanted=all&adxnnlx=1417608495-0Rfc8mF4vPBE/wDWxQiQ7Q.

———. (2013). Hackers in China Attacked The Times for Last 4 Months. Retrieved on October 5, 2014, from http://www.nytimes.com/2013/01/31/technology/chinese-hackers-infiltrate-new-york-times-computers.html?pagewanted=all&_r=0.

Pike, John. (2011). Vehicle-Mounted Active Denial System. Retrieved on February 25, 2015, from http://www.globalsecurity.org/military/systems/ground/v-mads.htm.

Pilger, Michael. (2015). Taiwan's Improving Patrol Fleet Could Enhance its Ability to Defend against a Chinese Invasion. Retrieved on April 20, 2015, from http://ori-gin.www.uscc.gov/sites/default/files/Research/Taiwan%20Improving%20Patrol%20Fleet_Staff%20Report_0.pdf.

Pillsbury, Michael. (2004). The US Role in Taiwan's Defense Reforms. Retrieved on April 20, 2015, from http://origin.www.uscc.gov/sites/default/files/Research/The%20US%20Role%20in%20Taiwan's%20Defense.pdf.

Poitras, Laura; Rosenbach, Marcel; and Stark, Holger. (2013). "Royal Concierge": GCHQ Monitors Diplomats' Hotel Bookings. Retrieved on October 25, 2014, from http://www.spiegel.de/international/europe/gchq-monitors-hotel-reservations-to-track-diplomats-a-933914.html.

Pollard, Niklas. (2014). Sweden Searches For Suspected "Foreign" Submarine Off Stockholm. Retrieved on November 21, 2014, from http://www.huffingtonpost.com/2014/10/18/sweden-search-submarine_n_6008686.html.

Poulsen, Kevin. (2009). Former Teen Hacker's Suicide Linked to TJX Probe. Retrieved on November 21, 2014, from http://www.wired.com/2009/07/hacker-3/.

President and deputy survive assassination attempt. (2004). Retrieved on April 25, 2015, from http://www.smh.com.au/articles/2004/03/20/1079199440157.html?from=storyrhs.

Prevent cyberterrorism. (2013). Retrieved on October 5, 2014, from http://news.xinhuanet.com/english/world/2013–12/27/c_132999955.htm.

Protalinski, Emil. (2012). Chinese spies used fake Facebook profile to friend NATO officials. Retrieved on October 2, 2014, from http://www.zdnet.com/blog/facebook/chinese-spies-used-fake-facebook-profile-to-friend-nato-officials/10389.

Qiang, Xiao. (2005). The Development and the State Control of the Chinese Internet. Retrieved on September 15, 2014, from http://www.uscc.gov/sites/default/files/4.14.05qiang_xiao_wrts.pdf.

Raduege, Harry D. Jr. (2004). Net-Centric Warfare Is Changing The Battlefield Environment. Retrieved on October 7, 2014, from http://www.crosstalkonline.org/stor-age/issue-archives/2004/200401/200401-Raduege.pdf.

Raff, Aviv. (2013). Chinese Time Bomb. Retrieved on October 13, 2014, from http://blog.seculert.com/2013/03/the-chinese-time-bomb.html.

Ragland, Leigh Ann; McReynolds, Joseph; Southerland, Matthew; and Mulvenon, James. (2013). Red Cloud Rising: Cloud Computing in China. Retrieved on November 23, 2014, from http://www.uscc.gov/sites/default/files/Research/DGI_Red%20Cloud%20Rising_2014.pdf.

Rawlinson, Kevin. (2015). US tech firms ask China to postpone "intrusive" rules. Retrieved on February 8, 2015, from http://www.bbc.com/news/technology-31039227.

Raz, Inbar. (2013). Physical (in)Security: It's Not All About Cyber. Retrieved on December 7, 2014, from https://www.youtube.com/watch?v=TbyNJ6fpd3U.

Recent Developments in China's Relations with Taiwan and North Korea. (2014). Retrieved on April 20, 2015, from http://origin.www.uscc.gov/sites/default/files/transcripts/Hearing%20Transcript_June%205,2014.pdf.

Reed, John. (2012a). Did Chinese Espionage Lead to F-35 Delays? Retrieved on October 5, 2014, from http://defensetech.org/2012/02/06/did-chinese-espionage-lead-to-f-35-delays/#more-16336.

———. (2012b). GAO Buys Fake Submarine Parts From China. Retrieved on October 2, 2014, from http://defensetech.org/2012/03/28/gao-buys-fake-submarine-parts-from-china/#more-16766.

———. (2012c). Proof That Military Chips From China Are Infected? Retrieved on October 5, 2014, from http://defensetech.org/2012/05/30/smoking-gun-proof-that-military-chips-from-china-are-infected/#more-17351.

Reid, Tim. (2007). China's cyber army is preparing to march on America, says Pentagon. Retrieved on October 2, 2014, from http://aftermathnews.wordpress.com/2007/09/08/china%E2%80%99s-cyber-army-is-preparing-to-march-on-america-says-pentagon/.

———. (2013). Facebook says target of sophisticated hacking attack. Retrieved on October 5, 2014, from http://www.reuters.com/article/2013/02/15/net-us-usa-social-facebook-idUSBRE91E16O20130215.

Reisinger, Don. (2012). Worldwide smartphone user base hits 1 billion. Retrieved on December 8, 2014, from http://www.cnet.com/news/worldwide-smartphone-user-base-hits-1-billion/.

Reporters Without Borders. (2005). The 11 Commandments of the Internet in China. Retrieved on September 21, 2014, from http://en.rsf.org/china-the-11-commandments-of-the-26–09–2005,15141.html.

Rhodes, Keith A. (2001). Information Security: Code Red, Code Red II, and SirCam Attacks Highlight Need for Proactive Measures. Retrieved on October 2, 2014, from http://www.gao.gov/new.items/d011073t.pdf.

Richardson, Michael. (2012). Thirst for energy driving China's foreign policy. Retrieved on April 7, 2015, from http://www.safpi.org/news/article/2012/thirst-energy-driving-chinas-foreign-policy.

Riley, Michael. (2015). Chinese Hackers Force Penn State to Unplug Engineering Computers. Retrieved on October 30, 2016, from http://www.bloomberg.com/news/articles/2015–05–15/china-hackers-force-penn-state-to-unplug-engineering-computers.

Riley, Michael, and Dune, Lawrence. (2012). Hackers Linked to China's Army Seen From EU to DC. Retrieved on October 13, 2014, from http://www.bloomberg.com/news/2012–07–26/china-hackers-hit-eu-point-man-and-d-c-with-byzantine-candor.html.

Riley, Michael, and Robertson, Jordan. (2015). China-Tied Hackers That Hit U.S. Said to Breach United Airlines. Retrieved on October 30, 2016, from http://www.bloomberg.com/news/articles/2015–07–29/china-tied-hackers-that-hit-u-s-said-to-breach-united-airlines.

Robertson, Ann E. (2011). *Militarization of Space*. New York: Facts on File.

Rogers, Mike. (2012). Investigative Report on the U.S. National Security Issues Posed by Chinese Telecommunications Companies Huawei and ZTE. Retrieved on October 2, 2014, from http://intelligence.house.gov/sites/intelligence.house.gov/files/documents/Huawei-ZTE%20Investigative%20Report%20(FINAL).pdf.

Rooker, J. W. (2008). Satellite Vulnerabilities. Retrieved on February 14, 2013, from http://www.dtic.mil/cgi-bin/GetTRDoc?AD=ADA507952.

Rumi, Aoyama. (2004). Chinese Diplomacy in the Multimedia Age: Public Diplomacy and Civil Diplomacy. Retrieved on September 23, 2014, from https://dspace.wul.waseda.ac.jp/dspace/bitstream/2065/800/1/20050307_aoyama_eng1.pdf.

Safire, William. (2004). The Farewell Dossier. Retrieved on December 2, 2014, from http://www.nytimes.com/2004/02/02/opinion/the-farewell-dossier.html.

Salmon, Diem. (2008). INEW China was a Cyber Threat. Retrieved on February 8, 2014, from http://dailysignal.com/2008/03/05/inew-china-was-a-cyber-threat/.

Sanger, David E., and Perlroth, Nicole. (2014). NSA Breached Chinese Servers Seen as Security Threat. Retrieved on November 21, 2014, from http://www.nytimes.com/2014/03/23/world/asia/nsa-breached-chinese-servers-seen-as-spy-peril.html.

Saporito, Laura, and Lewis, James A. (2013). Cyber Incidents Attributed to China. Retrieved on October 11, 2014, from http://csis.org/files/publication/130314_Chinese_hacking.pdf.

Schwartz, Felicia. (2015). Penn State's Engineering School Computers Hacked. Retrieved on October 30, 2016, from http://www.wsj.com/articles/penn-states-engineering-school-computers-hacked-1431804110.

Schwartz, John. (2007). When Computers Attack. Retrieved on October 2, 2014, from http://www.nytimes.com/2007/06/24/weekinreview/24schwartz.html?_r=1&oref=slogin.

Segal, Adam. (2012). Chinese Computer Games: Keeping Safe in Cyberspace. Retrieved on October 14, 2014, from http://www.foreignaffairs.com/articles/137244/adam-segal/chinese-computer-games.

———. (2014). Can China's New Internet Conference Compete with the West in Defining Norms of Cyberspace? Retrieved on September 21, 2015, from http://www.forbes.com/sites/adamsegal/2014/10/22/can-chinas-new-internet-conference-compete-with-the-west-in-defining-norms-of-cyberspace/.

Shambaugh, David. (2007). China's Propaganda System: Institutions, Processes and Efficacy. Retrieved on September 24, 2014, from http://myweb.rollins.edu/tlairson/china/chipropaganda.pdf.

Shanker, Thom. (2013). A New Medal Honors Drone Pilots and Computer Experts. Retrieved on October 5, 2014, from http://cryptome.org/2013/02/droners-hackers-killers.htm.

Skillings, Jonathan. (2012). Unmanned X-47B aircraft completes sea trial. Retrieved on February 26, 2013, from http://news.cnet.com/8301–11386_3–57560226–76/unmanned-x-47b-aircraft-completes-sea-trial/.

Southerland, Matthew. (2014a). Taiwan and China Agree to Enhance Communication, but Cross-Strait Economic Agreements Face Uncertainty. Retrieved on April 20, 2015, from http://origin.www.uscc.gov/sites/default/files/Research/Staff%20Report_Tai-wan%20and%20China%20Agree%20to%20Enhance%20Communication%20but%20Cross-Strait%20Economic%20Agreements%20Face%20Uncertainty.pdf.

———. (2014b). Taiwan's 2014 Local Elections: Implications for Cross-Strait Relations. Retrieved on April 20, 2015, from http://origin.www.uscc.gov/sites/default/files/Research/Staff%20Report_Taiwan's%20Local%20Elections--Implications%20for%20Cross-Strait%20Relations%20_12%2030%202014.pdf.

Space Security Index 2012. (2012). Retrieved on February 14, 2013, from http://swfound.org/media/93632/SSI_FullReport_2012.pdf.

Stokes, Mark A. (2010). Revolutionizing Taiwan's Security: Leveraging C4ISR for traditional and non-traditional challenges. Retrieved on April 20, 2015, from http://www.project2049.net/documents/revolutionizing_taiwans_security_leveraging_c4isr_for_traditional_and_non_traditional_challenges.pdf.

Stokes, Mark A., and Hsiao, Russell. (2012). Countering Chinese Cyber Operations: Opportunities and Challenges for U.S. Interests. Retrieved on October 13, 2014, from http://www2.gwu.edu/~nsarchiv/NSAEBB/NSAEBB424/docs/Cyber-079.pdf.

Stokes, Mark A., Lin, Jenny, and Hsiao, Russell. (2011). The Chinese People's Liberation Army Signals Intelligence and Cyber Reconnaissance Infrastructure. Retrieved on October 13, 2014, from http://project2049.net/documents/pla_third_department _sigint_cyber_stokes_lin_hsiao.pdf.

Storey, Ian. (2006). China's "Malacca Dilemma." Retrieved on April 7, 2015, from http:/ /www.jamestown.org/single/?no_cache=1&tx_ttnews%5Btt_news%5D=3943#.VS j3IWwcRH0.

Su Bin Complaint of Hacking for China. (2014). Retrieved on October 5, 2014, from http://cryptome.org/2014/08/su-bin-complaint-idg-news-14-0818.pdf.

Sun Tzu [Sun Zi]. (1963). *The Art of War* (trans. Samuel B. Griffith). London: Oxford University Press.

Taiwan-China: Recent Economic, Political and Military Developments across the Strait, and Implications for the United States. (2010). Retrieved on April 20, 2015, from http://origin.www.uscc.gov/sites/default/files/transcripts/3.18.10HearingTranscript.pdf.

Taiwan opposition party accuses China of hacking. (2011). Retrieved on October 5, 2014, from http://www.spacedaily.com/reports/Taiwan_opposition_party_accuses _China_of_hacking_999.html.

Tang, Didi. (2012). US social media account in China disappears. Retrieved on October 5, 2014, from http://news.yahoo.com/us-social-media-account-china-disappears-051645132--finance.html?_esi=1.

Targeting Huawei: NSA Spied on Chinese Government and Networking Firm. (2014). Retrieved on November 21, 2014, from http://www.spiegel.de/international/world/ nsa-spied-on-chinese-government-and-networking-firm-huawei-a-960199.html.

Taylor, Robert W., Fritsch, Eric J., and Liederbach, John. (2015). *Digital Crime and Digital Terrorism*. Upper Saddle River, NJ: Pearson, Inc.

Tham, Engen. (2015). China blasts NetEase for spreading porn in latest push to cleanse cyberspace. Retrieved on February 8, 2015, from http://www.reuters.com/article/ 2015/02/03/china-internet-netease-idUSL4N0VD1AY20150203.

The Internet in China. (2010). Information Office of the State Council of the People's Republic of China (Whitepaper). Retrieved on September 16, 2014, from http:// china.org.cn/government/whitepaper/node_7093508.htm.

The IP Commission Report (2013). Retrieved on October 18, 2014, from http:// www.ipcommission.org/report/ip_commission_report_052213.pdf.

The Organizational Structure of the State Council. (2003). Retrieved on September 24, 2014, from http://www.china.org.cn/english/kuaixun/64784.htm.

The State Development & Planning Commission of the People's Republic of China & Microsoft Corp. Sign A Memorandum of Understanding To Begin the Largest Joint Sino-Foreign Software Industry Cooperation. (2002). Retrieved on November 23, 2014, from http://news.microsoft.com/2002/06/27/the-state-development-planning-commission-of-the-peoples-republic-of-china-microsoft-corp-sign-a-memorandum-of-understanding-to-begin-the-largest-joint-sino-foreign-software-industry-coo/.

The United States Department of Justice. (2014). U.S. Charges Five Chinese Military Hackers for Cyber Espionage Against U.S. Corporations and a Labor Organization for Commercial Advantage. Retrieved on October 4, 2014, from http:// www.justice.gov/opa/pr/us-charges-five-chinese-military-hackers-cyber-espionage-against-us-corporations-and-labor.

Thomas, Timothy L. (2004). *Dragon Bytes*. Fort Leavenworth, KS: U.S. Government Printing Office.

———. (2005). Chinese and American Network Warfare. Retrieved on February 8, 2014, from http://library.uoregon.edu/ec/e-asia/read/netwar-1538.pdf.

———. (2007). *Decoding the Virtual Dragon*. Fort Leavenworth, KS: U.S. Government Printing Office.

———. (2009). *The Dragon's Quantum Leap*. Fort Leavenworth, KS: U.S. Government Printing Office.

Thornburgh, Nathan. (2005). Inside the Chinese Hack Attack. Retrieved on October 2, 2014, from http://www.time.com/time/nation/article/0,8599,1098371,00.html.

———. (2005). The Invasion of the Chinese Cyberspies. Retrieved on October 2, 2014, from http://www.time.com/time/magazine/article/0,9171,1098961,00.html.

Tiezzi, Shannon. (2014). Beijing's "China Threat" Theory. Retrieved on November 21, 2014, from http://thediplomat.com/2014/06/beijings-china-threat-theory/.

———. (2015). China (Finally) Admits to Hacking. Retrieved on April 20, 2015, from http://thediplomat.com/2015/03/china-finally-admits-to-hacking/.

Tkacik, John J. Jr. (2008). Trojan Dragons: China's Cyber Threat. Retrieved on October 7, 2014, from http://www.heritage.org/Research/asiaandthepacific/bg2106.cfm.

Tol, Jan van, Gunzinger, Mark, Krepinevich, Andrew F., and Thomas, Jim. (2010). AirSea Battle: A Point-Of-Departure Operational Concept. Retrieved on October 7, 2014, from http://www.csbaonline.org/publications/2010/05/airsea-battle-concept.

Trend Micro. (2014). Point-of-Sale System Breaches: Threats to the Retail and Hospitality Industries. Retrieved on December 6, 2014, from http://www.trend micro.com.au/cloud-content/us/pdfs/security-intelligence/white-papers/wp -pos-system-breaches.pdf.

Tung, Liam. (2007). China accused of cyberattacks on New Zealand. Retrieved on October 22, 2014, from http://news.cnet.com/China-accused-of-cyberattacks-on-New-Zealand/2100–7348_3–6207678.html.

UCS Satellite Database. (2012). Database, official names only. Retrieved on February 25, 2013, from http://www.ucsusa.org/nuclear_weapons_and_global_security/ space_weapons/technical_issues/ucs-satellite-database.html.

United States General Accounting Office. (2001). Information Security: Code Red, Code Red II, and SirCam Attacks Highlight Need for Proactive Measures. Retrieved on October 25, 2014, from http://www.gao.gov/new.items/d011073t.pdf.

U.S. Attorney's Office. (2014). Los Angeles Grand Jury Indicts Chinese National in Computer Hacking Scheme Allegedly Involving Theft of Trade Secrets. Retrieved on October 5, 2014, from http://www.fbi.gov/losangeles/press-releases/2014/los-angeles-grand-jury-indicts-chinese-national-in-computer-hacking-scheme-allegedly-involving-theft-of-trade-secrets.

U.S. Department of Defense. (2009). Annual Report to Congress: Military Power of the People's Republic of China 2009. Retrieved on September 15, 2014, from http://www.defense.gov/pubs/pdfs/China_Military_Power_Report_2009.pdf.

———. (2010). Annual Report to Congress: Military and Security Developments Involving the People's Republic of China 2010. Retrieved on September 15, 2014, from http://www.defense.gov/pubs/pdfs/2010_CMPR_Final.pdf.

———. (2011). Annual Report to Congress: Military and Security Developments Involving the People's Republic of China 2011. Retrieved on September 15, 2014, from http://www.defense.gov/pubs/pdfs/2011_CMPR_Final.pdf.

———. (2012). Annual Report to Congress: Military and Security Developments Involving the People's Republic of China 2012. Retrieved on September 15, 2014, from http://www.defense.gov/pubs/pdfs/2012_CMPR_Final.pdf.

———. (2013). Annual Report to Congress: Military and Security Developments Involving the People's Republic of China 2013. Retrieved on October 7, 2014, from http://www.defense.gov/pubs/2013_china_report_final.pdf.

———. (2014). Annual Report to Congress: Military and Security Developments Involving the People's Republic of China 2014. Retrieved on September 15, 2014, from http://www.defense.gov/pubs/2014_DoD_China_Report.pdf.

———. (2016). Annual Report to Congress: Military and Security Developments Involving the People's Republic of China 2016. Retrieved on October 30, 2016, from http://www.defense.gov/Portals/1/Documents/pubs/2016%20China%20Military%20Power%20Report.pdf.

US embassy cables: China uses access to Microsoft source code to help plot cyber warfare, US fears. (2010). Retrieved on November 23, 2014, from http://www .theguardian.com/world/us-embassy-cables-documents/214462?INTCMP=SRCH.

US hospital firm: Chinese hackers stole patient data. (2014). Retrieved on October 5, 2014, from http://www.spacedaily.com/reports/US_hospital_firm_Chinese_hackers _stole_patient_data_999.html.

US in new push to break China internet firewall. (2011). Retrieved on October 5, 2014, from http://www.smh.com.au/technology/security/us-in-new-push-to-break-china-internet-firewall-20110511–1ehze.html.

US Satellite Snaps China's First Aircraft Carrier At Sea. (2011). Retrieved on September 23, 2014, from http://www.theguardian.com/world/2011/dec/15/us-satellite-china-aircraft-carrier?newsfeed=true.

USCC Annual Report. (2002). Retrieved on September 22, 2014, from http://origin.www.uscc.gov/sites/default/files/annual_reports/2002%20Annual%20Report %20to%20Congress.pdf.

————. (2004). Retrieved on September 22, 2014, from http://origin.www.uscc.gov/ sites/default/files/annual_reports/2004-Report-to-Congress.pdf.

————. (2005). Retrieved on September 22, 2014, from http://origin.www.uscc.gov/ sites/default/files/annual_reports/2005-Report-to-Congress.pdf.

————. (2006). Retrieved on September 22, 2014, from http://origin.www.uscc.gov/ sites/default/files/annual_reports/USCC%20Annual%20Report%202006.pdf.

————. (2007). Retrieved on September 22, 2014, from http://origin.www.uscc.gov/ sites/default/files/annual_reports/2007-Report-to-Congress.pdf.

————. (2008). Retrieved on September 22, 2014, from http://origin.www.uscc.gov/ sites/default/files/annual_reports/2008-Report-to-Congress-_0.pdf.

————. (2009). Retrieved on September 22, 2014, from http://origin.www.uscc.gov/ sites/default/files/annual_reports/2009-Report-to-Congress.pdf.

————. (2010). Retrieved on September 22, 2014, from http://origin.www.uscc.gov/ sites/default/files/annual_reports/2010-Report-to-Congress.pdf.

————. (2011). Retrieved on September 22, 2014, from http://origin.www.uscc.gov/ sites/default/files/annual_reports/annual_report_full_11.pdf.

————. (2012). Retrieved on September 22, 2014, from http://origin.www.uscc.gov/ sites/default/files/annual_reports/2012-Report-to-Congress.pdf.

————. (2013). Retrieved on September 15, 2014, from http://origin.www.uscc.gov/ sites/default/files/annual_reports/Complete%202013%20Annual%20Report.PDF.

————. (2014). Retrieved on December 10, 2014, from http://origin.www.uscc.gov/ sites/default/files/annual_reports/Complete%20Report.PDF.

————. (2015). Retrieved on October 30, 2016, from http://origin.www.uscc.gov/sites/ default/files/annual_reports/2015%20Annual%20Report%20to%20Congress.PDF.

Verni, James. (2010). The Great Cyberheist. Retrieved on November 21, 2014, from http://www.nytimes.com/2010/11/14/magazine/14Hacker-t.html?pagewanted=all& _r=0.

Vijayan, Jaikumar. (2007). TJX data breach: At 45.6M card numbers, it's the biggest ever. Retrieved on December 1, 2014, from http://www.computerworld.com/article/ 2544306/security0/tjx-data-breach--at-45–6m-card-numbers--it-s-the-biggest-ever .html.

Villeneuve, Nart, Bennett, James T., Moran, Ned, Haq, Thoufique, Scott, Mike, and Geers, Kenneth. (2014). Operation "Ke3chang": Targeted Attacks Against Ministries of Foreign Affairs. Retrieved on October 25, 2014, from http://www.fireeye.com/ resources/pdfs/fireeye-operation-ke3chang.pdf.

Villeneuve, Nart, and Sancho, David. (2011). The "Lurid" Downloader. Retrieved on October 31, 2014, from http://la.trendmicro.com/media/misc/lurid-downloader-en-fal-report-en.pdf.

Vuving, Alexander L. (2014). Vietnam, the US, and Japan in the South China Sea. Retrieved on April 18, 2015, from http://thediplomat.com/2014/11/vietnam-the-us-and-japan-in-the-south-china-sea/.

Wakefield, Jane. (2013). China hackers "target EU foreign ministries." Retrieved on October 5, 2014, from http://www.bbc.com/news/technology-25316228.

Waterman, Shaun. (2007). China Has .75 Million Zombie Computers In US. Retrieved on March 17, 2008, from http://www.upi.com/International_Security/Emerging_Threats/Briefing/2007/09/17/china_has_75m_zombie_computers_in_us/7394/.

———. (2008). Chinese Cyberattacks Target US Think Tanks. Retrieved on October 2, 2014, from http://www.spacewar.com/reports/Chinese_Cyberattacks_Target_US_Think_Tanks_999.html.

Weinberger, Sharon. (2012). X-47B stealth drone targets new frontiers. Retrieved on February 26, 2013, from http://www.bbc.com/future/story/20121218-stealth-drone-targets-life-at-sea.

Weiss, Gus W. (2007). Duping the Soviets: The Farewell Dossier. Retrieved on December 2, 2014, from https://www.cia.gov/library/center-for-the-study-of-intelligence/kent-csi/vol39no5/pdf/v39i5a14p.pdf.

Weiss, Joseph. (2010). *Protecting Industrial Control Systems from Electronic Threats*. New York: Momentum Press.

Weitzenkorn, Ben. (2013). Internet Explorer Zero-Day Attack Targets Nuclear Researchers. Retrieved on October 5, 2014, from http://news.yahoo.com/internet-explorer-zero-day-attack-targets-nuclear-researchers-214704437.html.

White House says cyberattack thwarted. (2012). Retrieved on October 5, 2014, from http://www.usatoday.com/story/theoval/2012/10/01/obama-team-cyber-attack-thwarted/1605541/.

Williams, Carol J. (2014). Russia resuming Cold War-era bomber flights close to U.S. shores. Retrieved on November 21, 2014, from http://www.latimes.com/world/europe/la-fg-russia-bomber-flights-20141112-story.html.

Wilson, Clay. (2003). Computer Attack and Cyber Terrorism: Vulnerabilities and Policy Issues for Congress. Retrieved on December 2, 2014, from www.dtic.mil/cgi-bin/GetTRDoc?AD=ADA421056.

———. (2007). Network Centric Operations: Background and Oversight Issues for Congress. Retrieved on December 23, 2014, from http://www.fas.org/sgp/crs/natsec/RL32411.pdf.

———. (2008). Botnets, Cybercrime, and Cyberterrorism: Vulnerabilities and Policy Issues for Congress. Retrieved on December 2, 2014, from http://fas.org/sgp/crs/terror/RL32114.pdf.

Winkler, Ira. (2007). How To Take Down The Power Grid. Retrieved on December 8, 2014, from http://www.neardeathexperiments.com/smf/index.php?topic=2285.0.

Winkler, Tim. (2003). Dragonflies Prove Clever Predators. Retrieved on February 10, 2008, from http://info.anu.edu.au/ovc/media/Media_Releases/_2003/_030605Dragonflies.asp.

Wines, Michael. (2011). China Creates New Agency for Patrolling the Internet. Retrieved on September 24, 2014, from http://www.nytimes.com/2011/05/05/world/asia/05china.html?_r=1&.

Winn, Patrick. (2008). Hypothetical attack on U.S. outlined by China. Retrieved on December 2, 2014, from http://www.airforcetimes.com/news/2008/01/airforce_china_strategy_080121/.

Wiseman, Len. (2007). *Live Free or Die Hard*. Los Angeles, CA: 20th Century Fox.

Wolchok, Scott, Yao, Randy, and Halderman, Alex J. (2009). Analysis of the Green Dam Censorware System. Retrieved on October 25, 2014, from https://jhalderm.com/pub/gd/.

Wolchover, Natalie. (2011). Google Maps Mystery Actually Spy Satellite Targets, Expert Says. Retrieved on September 23, 2014, from http://www.foxnews.com/tech/2011/11/17/mysterious-symbols-in-china-desert-are-spy-satellite-targets-expert-says/.

Wong, Edward. (2013). On Scale of 0 to 500, Beijing's Air Quality Tops "Crazy Bad" at 755. Retrieved on October 7, 2014, from http://www.nytimes.com/2013/01/13/science/earth/beijing-air-pollution-off-the-charts.html?_r=0.

Wong, Wilson W. S., and Fergusson, James. (2010). *Military Space Power*. Santa Barbara, CA: Praeger.

Wortzel, Larry M. (2013). *The Dragon Extends Its Reach*. Washington, DC: Potomac Books.

——. (2014). The Chinese People's Liberation Army and Information Warfare. Retrieved on April 26, 2015, from http://www.strategicstudiesinstitute.army.mil/pdffiles/PUB1191.pdf.

Wu, Jiao. (2006). Nanjing pays tribute to "Conscience of Japan." Retrieved on April 20, 2015, from http://www.chinadaily.com.cn/english/doc/2006–01/06/content_509646.htm.

Xi Jinping leads Internet security group. (2014). Retrieved on October 5, 2014, from http://news.xinhuanet.com/english/china/2014–02/27/c_133148273.htm?utm_source=The+Sinocism+China+Newsletter&utm_campaign=d96724e196-Sinocism02_27_14.

Yahoo Implicated In Third Cyberdissident Trial. (2006). Retrieved on September 22, 2014, from http://en.rsf.org/china-yahoo-implicated-in-third-19–04–2006,17180.html.

Yamei, Wang. (2014). Death sentences upheld in Kunming terrorist attack case. Retrieved on December 1, 2014, from http://news.xinhuanet.com/english/china/2014–10/31/c_133756513.htm.

Yue, Qi, and Yue, Qin. (2008). China Regime Implicated In Staging Violence in Tibetan Protest. Retrieved on September 23, 2014, from http://chinaview.wordpress.com/2008/03/29/photo-china-regime-implicated-in-staging-violence-in-tibet-protest/.

Index

About the Author

Jason R. Fritz earned his PhD from Bond University in Australia, including field research conducted in China and Israel. He earned his undergraduate degree in the United States and has studied and traveled across five continents, including study at China's Tianjin Foreign Studies University. His publications on cyber warfare terminology, satellite security, nuclear command and control, and Chinese cyber warfare have garnered widespread online attention. Jason's university teaching experience includes the subjects of geopolitics, globalization, Chinese strategy, and Chinese defense policy. His interest in computers dates to the Apple II and Commodore 64 of the early 1980s.

www.ingramcontent.com/pod-product-compliance
Lightning Source LLC
Chambersburg PA
CBHW051235050326
40689CB00007B/929

9 781498 537094